Transition Metal Catalyzed Furan Synthesis

Transition Metal Catalyzed Furan Synthesis

Transition Metal Catalyzed Heterocycles Synthesis Series

Xiao-Feng Wu
Department of Chemistry, Zhejiang Sci-Tech University,
Hangzhou, People's Republic of China
and
Leibniz-Institut für Katalyse e.V., Universität Rostock,
Rostock, Germany

ELSEVIER

AMSTERDAM • BOSTON • HEIDELBERG • LONDON
NEW YORK • OXFORD • PARIS • SAN DIEGO
SAN FRANCISCO • SINGAPORE • SYDNEY • TOKYO

Elsevier
Radarweg 29, PO Box 211, 1000 AE Amsterdam, Netherlands
The Boulevard, Langford Lane, Kidlington, Oxford OX5 1GB, UK
225 Wyman Street, Waltham, MA 02451, USA

Notices
Knowledge and best practice in this field are constantly changing. As new research and
experience broaden our understanding, changes in research methods or professional practices,
may become necessary.

Practitioners and researchers must always rely on their own experience and knowledge in
evaluating and using any information or methods described herein. In using such information or
methods they should be mindful of their own safety and the safety of others, including parties for
whom they have a professional responsibility.

To the fullest extent of the law, neither the Publisher nor the authors, contributors, or editors,
assume any liability for any injury and/or damage to persons or property as a matter of products
liability, negligence or otherwise, or from any use or operation of any methods, products,
instructions, or ideas contained in the material herein.

ISBN: 978-0-12-804034-8

British Library Cataloguing-in-Publication Data
A catalogue record for this book is available from the British Library

Library of Congress Cataloging-in-Publication Data
A catalog record for this book is available from the Library of Congress

For Information on all Elsevier publications
visit our website at http://store.elsevier.com/

This book has been manufactured using Print On Demand technology.

DEDICATION

Dedicated to my wife, children, and my parents

Thanks for their understanding, support, encouragement, and tolerance!

CONTENTS

CONTENTS

CHAPTER 1

Introduction

Furan is a class of five-membered heterocyclic organic compounds, consisting of four carbon atoms and one oxygen atom with two double bonds in the aromatic ring. Until today, numerous studies have been reported on the applications of compounds with furan as the core moiety, such as in advanced materials, natural products, pharmaticials, etc. (Scheme 1.1). Based on these importances, methodologies have been developed for the selective and effective preparation of diversely substituted furan cores during the past few decades [1]. In this volume, the progresses of transition metal-catalyzed furan synthesis will be discussed generally. Based on the substrates applied, the whole volume is catalogued by [2 + 2 + 1] cyclizations, [3 + 2] cyclizations, [4 + 1] cyclizations, and intramolecular cyclization reactions.

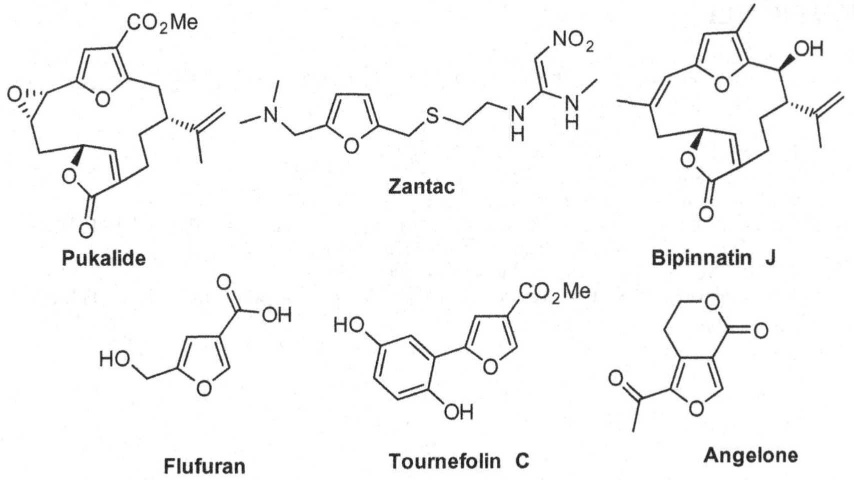

Scheme 1.1 Selected examples of bio-active furan derivatives.

Transition Metal Catalyzed Furan Synthesis. DOI: http://dx.doi.org/10.1016/B978-0-12-804034-8.00001-5

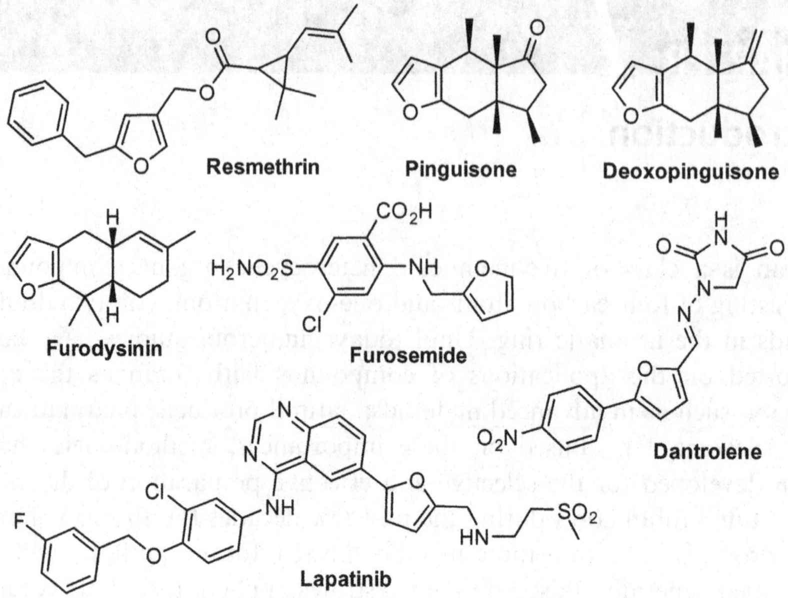

Resmethrin   **Pinguisone**   **Deoxopinguisone**

Furodysinin   **Furosemide**   **Dantrolene**

Lapatinib

Scheme 1.1 (Continued)

REFERENCE

[1] For synthetic reviews on furan synthesis see:

a. X.L. Hou, H.Y. Cheung, T.Y. Hon, P.L. Kwan, T.H. Lo, S.Y. Tong, et al., Tetrahedron 54 (1998) 1955–2020.
b. M. Egi, S. Akai, Heterocycles 91 (2015) 931–958.
c. N.T. Patil, Y. Yamamoto, Arkivoc x (2007) 121–141.
d. B. Crone, S.F. Kirsch, Chem. Eur. J. 14 (2008) 3514–3522.
e. C. Song, J. Wang, Z. Xu, Org. Biomol. Chem. 12 (2014) 5802–5806.
f. T.J. Donohoe, J.F. Bower, L.K.M. Chan, Org. Biomol. Chem. 10 (2012) 1322–1328.
g. S.F. Kirsch, Org. Biomol. Chem. 4 (2006) 2076–2080.
h. R.C.D. Brown, Angew. Chem. Int. Ed. 44 (2005) 850–852.
i. A.V. Gulevich, A.S. Dudnik, N. Chernyak, V. Gevorgyan, Chem. Rev. 113 (2013) 3084–3213.

CHAPTER 2

Synthesized by [2 + 2 + 1] Cyclization Reactions

For the [2 + 2 + 1] cyclization reactions developed for furan synthesis, alkynes or their surrogates are the commonly applied substrates. In 2009, Jiang and co-workers developed a novel method for the synthesis of tetrasubstituted furans from aromatic alkynes [1]. In this procedure, the presence of Lewis acid and oxygen is essential because good yields of the furans can be obtained in the presence of $Zn(OTf)_2$ (0.3 equiv.) and O_2 (7.6 bar). Here, the formation of 1,2,3,4-tetraphenylbut-2-ene-1,4-dione from two alkynes was proposed and confirmed as the key intermediate and followed by intramolecular cyclization to give the final furans (Scheme 2.1). Notably, by using $K_2S_2O_8$ as the oxidant and in the absence of Lewis acid, the corresponding cyclopentadienyl alkyl ethers can be produced in good yields [2].

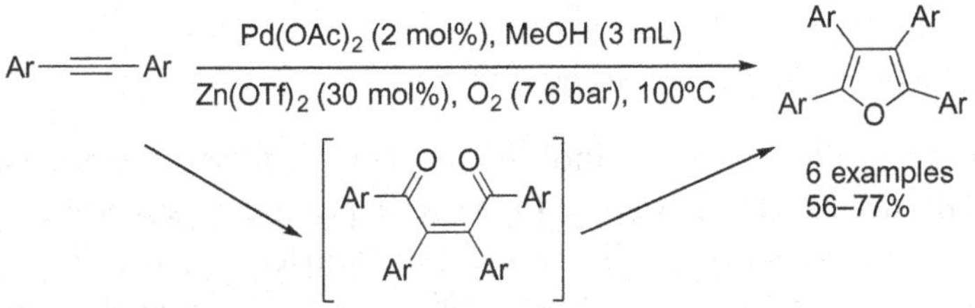

Scheme 2.1 Palladium-catalyzed synthesis of furans from alkynes.

Later on, the same group improved this transformation and realized the transformation running under atmospheric pressure of oxygen (Scheme 2.2) [3]. The reaction runs in a fluorous biphasic system of N,N-dimethylacetamide and perfluorodecalin at 60 °C, and the corresponding tetrasubstituted furans were isolated in good yields in general. Not only intermolecular but also intramolecular alkyne–alkyne cyclization can be realized under similar reaction conditions. However, compared with primary case, higher catalyst loading is required.

Transition Metal Catalyzed Furan Synthesis. DOI: http://dx.doi.org/10.1016/B978-0-12-804034-8.00002-7

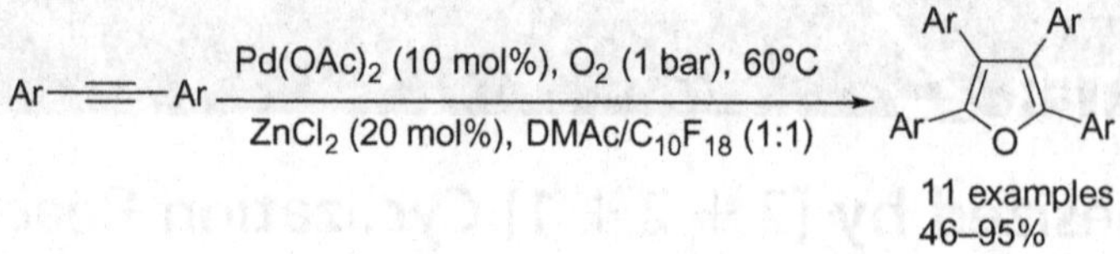

$$\text{Ar}\!\!-\!\!\!\equiv\!\!\!-\!\!\text{Ar} \xrightarrow[\text{ZnCl}_2 \ (20 \ \text{mol}\%), \ \text{DMAc/C}_{10}\text{F}_{18} \ (1{:}1)]{\text{Pd(OAc)}_2 \ (10 \ \text{mol}\%), \ \text{O}_2 \ (1 \ \text{bar}), \ 60^\circ\text{C}}$$

11 examples
46–95%

Scheme 2.2 Palladium-catalyzed furans from alkynes in biphasic system.

Recently, a PdCl$_2$-catalyzed procedure has been reported [4]. With dioxygen as the sole oxidant in N,N-dimethylacetamide-water, tetrasubstituted furans were produced in moderate to good yields (Scheme 2.3). Aryl halides were compatible and the oxygen atom of the furan was from water instead of molecular oxygen. From the mechanistic point of view, the authors proposed the formation of vinylpalladium species which is different with previous cases.

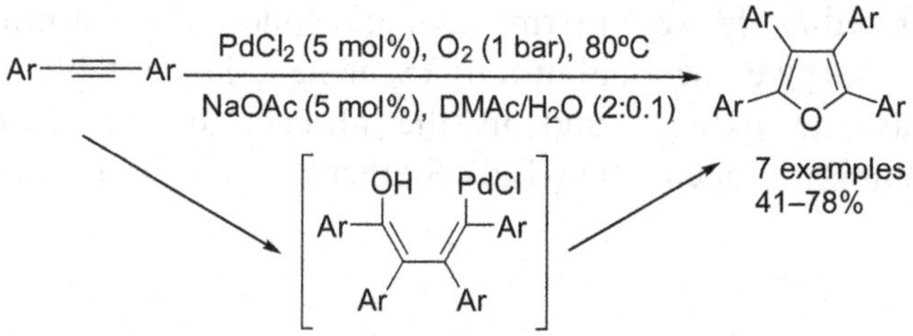

$$\text{Ar}\!\!-\!\!\!\equiv\!\!\!-\!\!\text{Ar} \xrightarrow[\text{NaOAc} \ (5 \ \text{mol}\%), \ \text{DMAc/H}_2\text{O} \ (2{:}0.1)]{\text{PdCl}_2 \ (5 \ \text{mol}\%), \ \text{O}_2 \ (1 \ \text{bar}), \ 80^\circ\text{C}}$$

7 examples
41–78%

Scheme 2.3 Palladium-catalyzed furans from alkynes in water.

More recently, Zhao and co-workers further improved this transformation and reported a facile and practical procedure for the synthesis of tetrasubstituted furans [5]. Started from alkynes with palladium acetate as the catalyst together with cupric acetate in acetic acid, under atmospheric of oxygen, various target furans were formed in satisfactory yields (Scheme 2.4). Notably, this reaction was shown be possible even if it is performed at gram scale as well.

Scheme 2.4

Alkyne (1 mmol), Pd(OAc)$_2$ (5 mol%), Cu(OAc)$_2$ (10 mol%), and AcOH (2 ml) were added into a test tube attached to an oxygen balloon (1 bar). The system was stirred magnetically and heated at 100 °C; with an oil bath for 12 h. And then the reaction was quenched by the addition of 10 ml water. The aqueous solution was extracted with ethyl acetate (10 ml × 3) and the combined extract was dried with anhydrous MgSO$_4$. The solvent was vacuumed and the crude product was isolated by TLC with light petroleum ether/DCM as eluent to give the pure product.

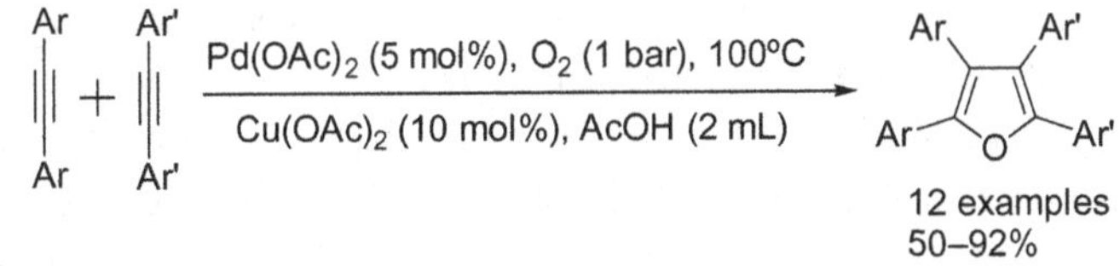

Ar Ar'
‖ + ‖   Pd(OAc)$_2$ (5 mol%), O$_2$ (1 bar), 100°C
Ar Ar'   ⎯⎯⎯⎯⎯⎯⎯⎯⎯⎯⎯⎯⎯⎯⎯⎯⎯⎯⎯⎯⎯→
         Cu(OAc)$_2$ (10 mol%), AcOH (2 mL)

12 examples
50–92%

Scheme 2.4 Palladium-catalyzed furans from alkynes in AcOH.

In 2015, Guo's group developed a gold-catalyzed approach for the synthesis of 2,5-disubstituted furans [6]. By using PPh$_3$Au(I)OTf as the catalyst and haloalkynes as the substrates, good yields of the furan derivatives can be produced (Scheme 2.5a). Regarding the reaction mechanism, the reaction started with the activation of the bromoalkyne to give 1,3-diyne under the assistant of PPh$_3$Au(I)OTf. Then the 1,3-diyne undergoes nucleophilic addition of H$_2$O with the help of base and gold catalyst to give the hydration product which gave the final product after intramolecular 5-exodig cyclization. Importantly, the presence of water is compulsory here because traces of the desired product can be detected only in the absence of water. This transformation was reported with copper as the catalyst as well [7]. In the presence of CuI (5 mol%), 1,10-phen (15 mol%), KOH in DMSO, good yields of the desired furans were formed at 80°C. During the same period, Lee and co-workers developed a one-pot synthesis of 2,5-diaryl-substituted furans from the homo-coupling and cyclization of aryl alkynyl carboxylic acids [8]. Aryl alkynyl carboxylic acids are stable substrates that can be easily prepared from the coupling reaction of propiolic acid and aryl halides. 1,4-Diaryl-1,3-butadiynes were found to be the intermediates and not needed to be isolated. A large class of furans was prepared in good yields (Scheme 2.5b). In addition, thiophenes could be synthesized under the optimized conditions via the sequential addition of Na$_2$S · 9H$_2$O.

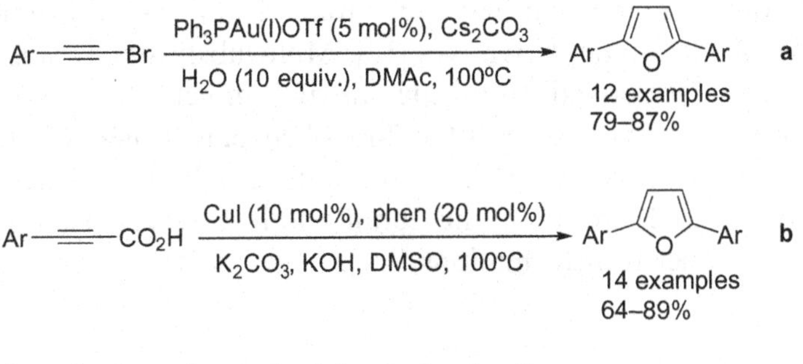

Ar══Br   Ph$_3$PAu(I)OTf (5 mol%), Cs$_2$CO$_3$
          ⎯⎯⎯⎯⎯⎯⎯⎯⎯⎯⎯⎯⎯⎯⎯⎯⎯⎯⎯⎯⎯→   Ar⌝O⌐Ar   a
          H$_2$O (10 equiv.), DMAc, 100°C

12 examples
79–87%

Ar══CO$_2$H   CuI (10 mol%), phen (20 mol%)
               ⎯⎯⎯⎯⎯⎯⎯⎯⎯⎯⎯⎯⎯⎯⎯⎯⎯⎯⎯⎯→   Ar⌝O⌐Ar   b
               K$_2$CO$_3$, KOH, DMSO, 100°C

14 examples
64–89%

Scheme 2.5 Furans from bromoalkynes and aryl alkynyl carboxylic acids.

In 1992, Takai, Utimoto and their co-workers developed a tantalum-mediated procedure for the synthesis of furans [9]. A variety of 2,3,4-trisubstituted furans can be prepared by the treatment of tantalum—alkyne complexes with aldehydes, followed by addition of an isocyanide in DME-PhH-THF(1:1:1). In this procedure, the tantalum—alkyne complexes were generated *in situ* by combining alkynes, TaCl$_5$, and Zn in DME and benzene. Moderate yields of the products can be achieved (Scheme 2.6a). Later on, Urabe, Sato and their research groups reported the preparation of furans via azatitanacyclopentadiene intermediates which produced selectively from acetylene, nitrile, and a titanium reagent {Ti(O-i-Pr)$_4$/i-PrMgCl} [10]. Moderate yields of the desired furans can be produced here (Scheme 2.6b).

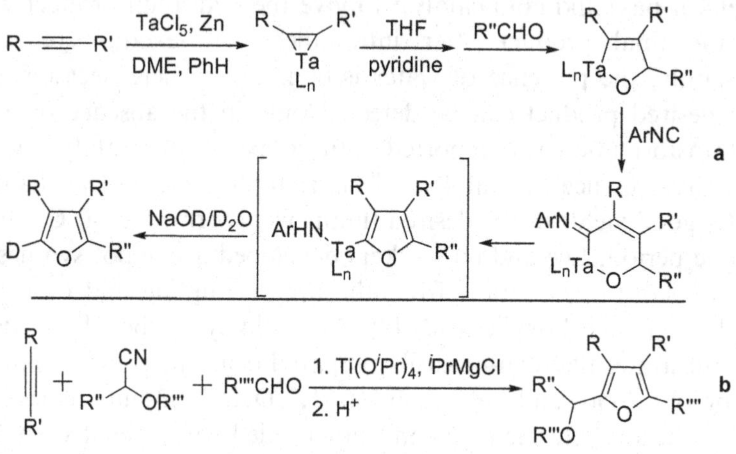

Scheme 2.6 Ta/Ti-mediated furans synthesis.

In 2014, Yoshikai and co-workers reported a palladium-catalyzed methodology for the synthesis of furans from *N*-aryl imines and alkynylbenziodoxolone derivatives [11]. Multisubstituted furans, whose substituents are derived from the alkynyl moiety (2-position), the imine (3- and 4-positions), and the 2-iodobenzoate moiety (5-position), were produced in good yields under mild conditions (Scheme 2.7). The 2-iodophenyl group of the furan product serves as a versatile handle for further transformations.

Scheme 2.7

In a Schlenk tube were placed Pd(OAc)$_2$ (4.4 mg, 0.020 mmol), 1-(phenylethynyl)-1,2-benziodoxol-3(1H)-one (139 mg, 0.40 mmol), and (E)-N-(1-(biphenyl-4-yl)ethylidene)-4-methoxyaniline (60 mg, 0.20 mmol) under nitrogen atmosphere, followed by the addition of toluene (1 ml). The resulting mixture was stirred at room temperature for 24 h. The solvent was evaporated under reduced pressure to afford the crude product. The crude product was purified by silica gel chromatography (eluent: hexane/EtOAc = 100/1) to afford the pure product.

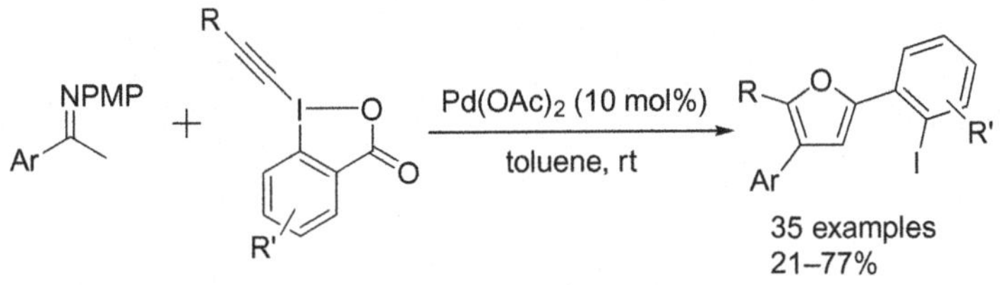

Scheme 2.7 Palladium-catalyzed synthesis of multisubstituted furans.

REFERENCES

[1] A. Wang, H. Jiang, Q. Xu, Synlett (2009) 929–932.

[2] R. Cai, M. Huang, X. Cui, J. Zhang, C. Du, Y. Wu, et al., RSC Adv. 3 (2013) 13140–13143.

[3] Y. Wen, S. Zhu, H. Jiang, A. Wang, Z. Chen, Synlett (2011) 1023–1027.

[4] L. Wang, J. Li, Y. Lv, G. Zhao, S. Gao, Synlett 23 (2012) 1074–1078.

[5] J. Xu, X. Song, J. Zhao, Chin. J. Chem. 32 (2014) 1099–1102.

[6] P. Guo, Catal. Commun. 68 (2015) 58–60.

[7] H. Jiang, W. Zeng, Y. Li, W. Wu, L. Huang, W. Fu, J. Org. Chem. 77 (2012) 5179–5183.

[8] F.M. Irudayanathan, G.C.E. Raja, S. Lee, Tetrahedron 71 (2015) 4418–4425.

[9] Y. Kataoka, M. Tezuka, K. Takai, K. Utimoto, Tetrahedron 48 (1992) 3495–3502.

[10] D. Suzuki, Y. Nobe, Y. Watai, R. Tanaka, Y. Takayama, F. Sato, et al., J. Am. Chem. Soc. 127 (2005) 7474–7479.

[11] B. Lu, J. Wu, N. Yoshikai, J. Am. Chem. Soc. 136 (2014) 11598–11601.

CHAPTER 3

Synthesized by [3 + 2] Cyclization Reactions

In the reported [3 + 2] cyclization for furan synthesis, 1,3-diketones or propargyl alcohol derivatives are usually applied as the core component. In 1985, Tsuji and co-workers reported a palladium-catalyzed cyclization of propargyl carbonates with 1,3-diketones to produce the corresponding 4-methylfurans [1]. In the presence of palladium catalyst under neutral conditions, good yields of the desired furans can be isolated (Scheme 3.1). For the reaction mechanism, the generation of 1,2-propadienylpalladium carbonate complex was proposed.

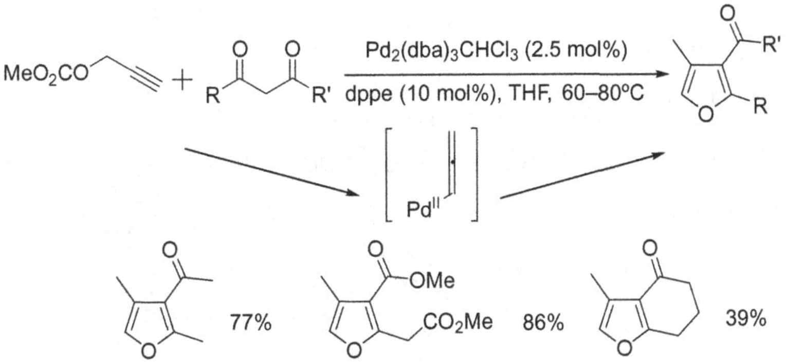

Scheme 3.1 Palladium-catalyzed synthesis of furans under neutral conditions.

Trost and McIntosh reported a tandem process for the synthesis of furans in 1995 [2]. Terminal alkynes and γ-hydroxy ynoates were applied as the substrates. Depending on γ-hydroxy ynoates applied, furans or butenolides can be selectively produced.

Cadierno, Gimeno, and their co-workers developed a ruthenium-catalyzed simple and efficient procedure for the preparation of tetrasubstituted furans [3]. Starting from readily accessible propargylic alcohols and commercially available 1,3-dicarbonyl compounds, good yields of the expected furans were formed (Scheme 3.2a). The reaction proceeded in a one-pot manner, involves the initial trifluoroacetic

Transition Metal Catalyzed Furan Synthesis. DOI: http://dx.doi.org/10.1016/B978-0-12-804034-8.00003-9

acid-promoted propargylation of the 1,3-dicarbonyl compound, and subsequent allyl-ruthenium(II) complex [Ru(η^3-2-C$_3$H$_4$Me)(CO)(dppf)] [SbF$_6$]-catalyzed cycloisomerization of the γ-ketoalkyne. In the substrates testing, they found that 6,7-dihydro-5H-benzofuran-4-ones can be formed when 1,3-cyclohexanediones were applied as the coupling partner. Soon later, Zhan and co-workers explored the possibility of applying base metal as the catalyst. With FeCl$_3$ [4] or Cu(OTf)$_2$ [5] as the catalyst, propargylic alcohols or acetates and 1,3-dicarbonyl compounds or enoxysilanes as the substrates, the desired substituted furans can be formed in good yields (Scheme 3.2b).

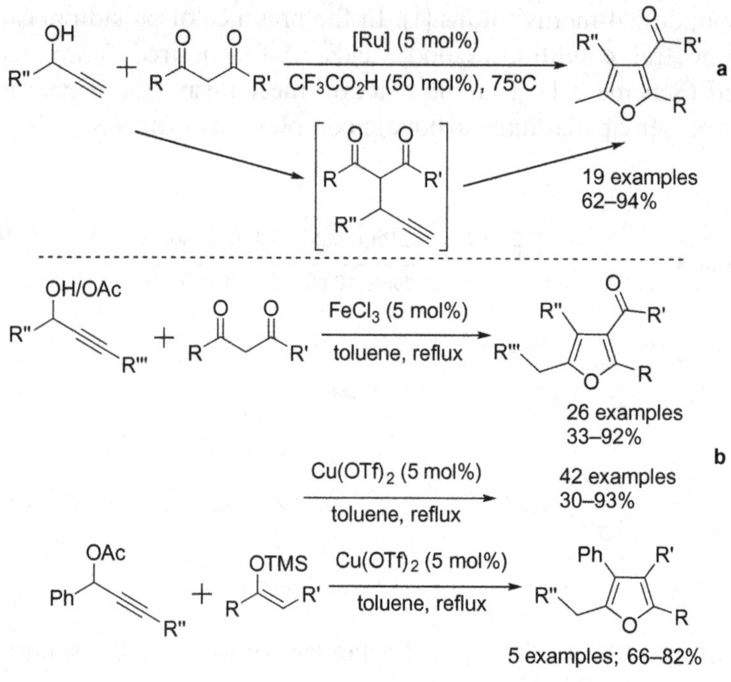

Scheme 3.2 Ru/Fe/Cu-catalyzed synthesis of furans.

Scheme 3.2b

To a 5 mL flask, propargylic alcohols or propargylic acetates (0.5 mmol), 1,3-dicarbonyl compounds (2.0 mmol), toluene (2.0 mL), and FeCl$_3$ (0.025 mmol, 4 mg) were successively added. The reaction mixture was stirred at reflux and monitored periodically by TLC. Upon completion, toluene was removed under reduced pressure by an aspirator, and then the residue was purified by silica gel column chromatography (EtOAc−hexane) to give the pure product.

To a 5 mL flask, 1-(naphthalen-1-yl)-3-(trimethylsilyl)prop-2-yn-1-ol (127 mg, 0.5 mmol), ethyl acetoacetate (195 mg, 1.5 mmol), toluene (2.0 mL), and Cu(OTf)$_2$ (9 mg, 0.025 mmol) were successively added. The reaction mixture was stirred at reflux and monitored periodically by TLC. Upon completion, the toluene was removed under reduced pressure by an aspirator, and then the residue was purified by silica gel column chromatography (EtOAc/ hexane) to afford the pure product.

In 2003, a novel ruthenium- and platinum-catalyzed sequential reaction to afford the corresponding tri- and tetrasubstituted furans or pyrroles was reported [6]. Starting from propargylic alcohols with ketones, or with ketones and anilines, furans or pyrroles were formed in moderate to good yields with high regioselectivities (Scheme 3.3). For the reaction pathway, the reaction started with transforming propargylic alcohols and ketones into γ-ketoalkyne by ruthenium catalyst. Then, followed by PtCl$_2$-catalyzed hydration of the alkyne moiety by the produced H$_2$O slowly gives the 1,4-diketone which will produce the desired furans after intramolecular cyclization.

Scheme 3.3

[Cp*RuCl(μ$_2$-SMe)$_2$-RuCp*Cl] (38 mg, 0.06 mmol), NH$_4$BF$_4$ (12 mg, 0.12 mmol), and PtCl$_2$ (31 mg, 0.12 mmol) were placed in a 50 mL flask under N$_2$. Anhydrous acetone (30 mL) was added, and then the mixture was magnetically stirred at room temperature. After the addition of propargylic alcohol (0.60 mmol), the reaction flask was kept at reflux temperature for 36 h. The solvent was concentrated under reduced pressure by an aspirator, and then the residue was purified by TLC (SiO$_2$) with EtOAc-hexane (1/9) to give the pure product.

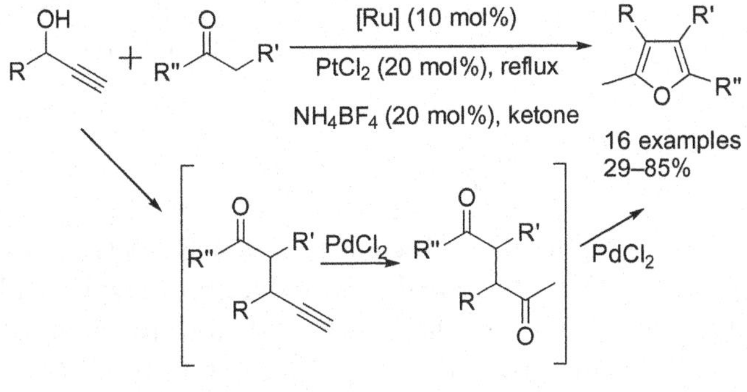

Scheme 3.3 Ru and Pt-catalyzed synthesis of furans.

Willis and co-workers demonstrated a rhodium-catalyzed intermolecular hydroacylation and applied to the synthesis of di- and trisubstituted furans in 2011 [7]. The reaction proceeds through hydroacylation and followed by a simple dehydrative cyclization; moderate to good yields of the desired furans can be isolated (Scheme 3.4).

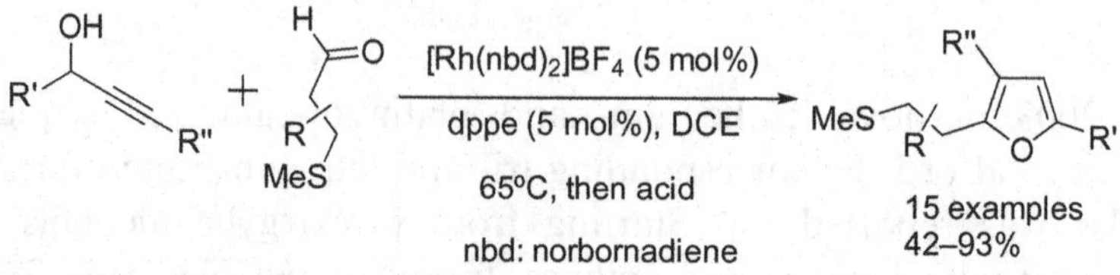

Scheme 3.4 *Rh-catalyzed synthesis of furans* via *hydroacylation.*

Bach and Nitsch reported a methodology for the synthesis of 2-alkenylfurans in 2014 [8]. The reaction relies on a successive twofold S_N'-type substitution reaction at methoxy-substituted propargylic acetates. The furan C3 − C4 bond is presumably established by silyl enol ether attack at a propargylic cation intermediate. The resulting α-methoxyallene is intramolecularly substituted, leading to cyclization by displacement of the methoxy group (O − C2 bond formation) and to simultaneous formation of the exocyclic alkene double bond. With bismuth(III) triflate as the catalyst, a variety of 2-alkenylfurans were isolated in good yields (Scheme 3.5).

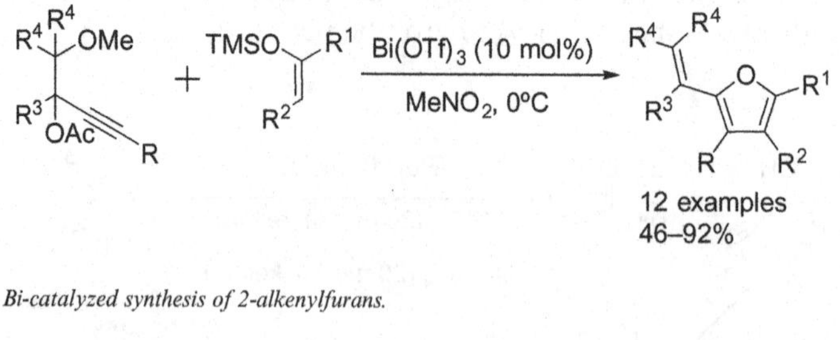

Scheme 3.5 *Bi-catalyzed synthesis of 2-alkenylfurans.*

Balme's group reported an efficient one-step synthesis of furofurans and furopyrroles from the easily available propargyl alcohols (or amines) and arylidene (or alkylidene) β-ketosulfones in 2000 [9]. Moderate yields of the furans can be produced (Scheme 3.6).

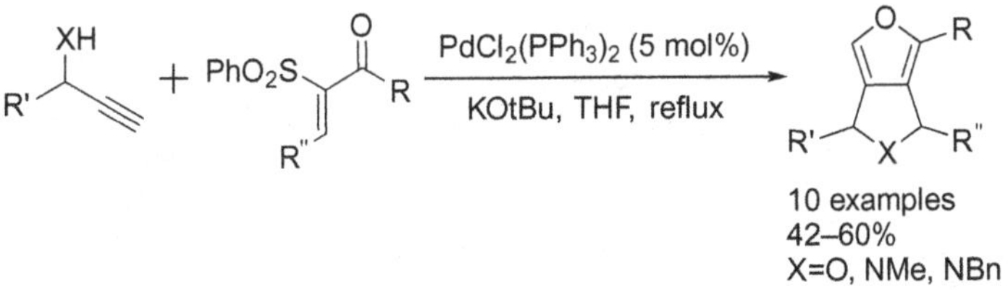

Scheme 3.6 Pd-catalyzed synthesis of furofurans.

In 2003, a palladium-catalyzed reaction of acylchromates and propargylic tosylates was developed [10]. Substituted furans are prepared in moderate to good yields (Scheme 3.7). From mechanistic point of view, the reaction was initiated by the oxidative addition of propagylic tosylates to palladium(0) complexes to give 1,2-propadienylpalladium(II) complexes, which then reacted with acylchromates to form 1,2-propadienyl ketones. And then the 1,2-propadienyl ketones were transformed to furans by the action of *in situ* generated $Cr(CO)_5$.

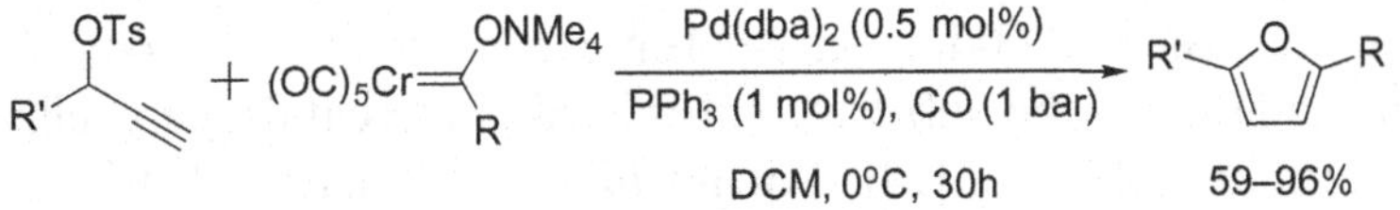

Scheme 3.7 Pd-catalyzed synthesis of furans from acylchromates.

In 2009, Jiang and co-workers developed a novel and efficient method for the regiospecific synthesis of polysubstituted furan aldehydes/ketones [11]. The reaction proceeded *via* a copper(I)-catalyzed rearrangement/dehydrogenation oxidation/carbene oxidation sequence of 1,5-enynes which *in situ* formed from alkynols and diethyl but-2-ynedioate under atmospheric pressure. Highly functionalized furans were produced in moderate to good yields (Scheme 3.8). Later on, they found nano-Cu_2O can be the catalyst as well. With the same substrates, different selectivity can be observed when silver was applied as the catalyst. An interesting methodology for the synthesis of 3-alkylidene furans from α-alkynyl epoxides and β-keto esters with palladium as the catalyst was reported as well [12].

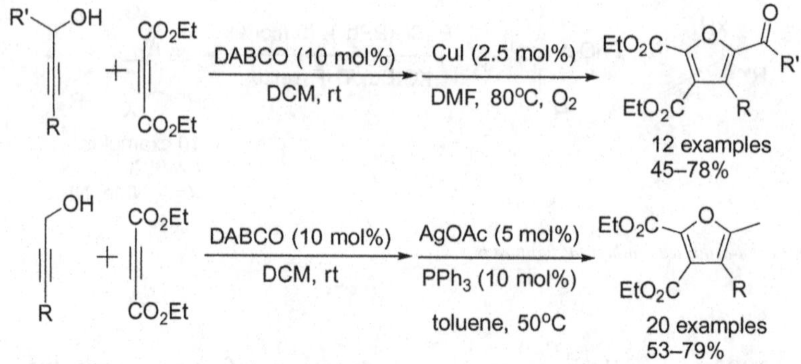

Scheme 3.8 Cu-catalyzed synthesis of polysubstituted furans.

In 1994, Pirrung and co-workers developed a rhodium-catalyzed dipolar cycloaddition of cyclic rhodium carbenoids to diagonal carbon for the synthesis of furans [13]. By using diazocyclohexanediones and terminal alkynes as the substrates, the corresponding furans were formed in moderate to good yields (Scheme 3.9). Interestingly, fluoro-benzene was applied as the solvent and reactions were run at room temperature. Later on, they further explored this transformation and succeeded in extending the substrates to terminal alkenes. Lee's group studied the using of 3-diazo-2,4-chromenediones and 2-diazo-1H-1,3-phenalenedione as the substrates in this rhodium-catalyzed cyclization with terminal alkynes [14]. Meanwhile, isocyanates, ketones, and olefins were applied as coupling partners as well.

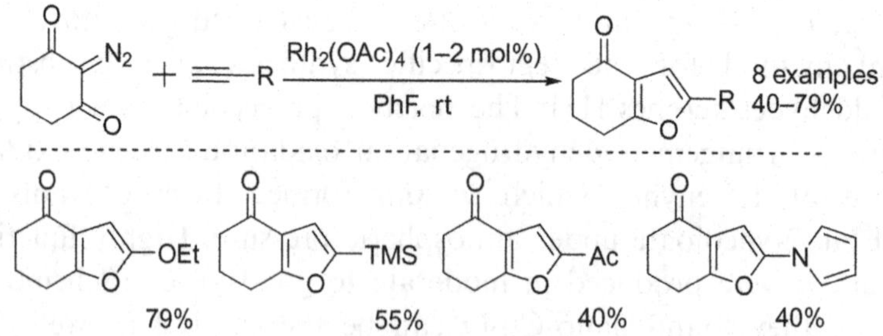

Scheme 3.9 Rh-catalyzed synthesis of furans from diazocyclohexanediones.

In 2009, a $Rh_2(OAc)_4$-catalyzed process for the cyclopropenations of ynamides was developed by Hsung and Li [15]. Highly substituted

2-amido-furans were formed in good yields by this formerly [3 + 2] cycloaddition. Interestingly, the products can go [4 + 2] cycloadditions to give the corresponding dihydroindoles and tetrahydroquinolines under thermal conditions (Scheme 3.10). In this procedure, in addition to diazo malonate and ethyl α-diazoacetate, the corresponding phenyl iodonium ylide was tested as well.

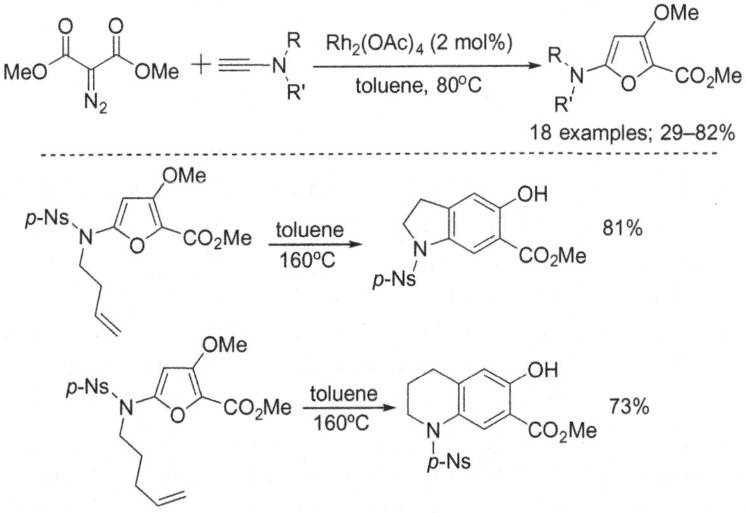

Scheme 3.10 Rh-catalyzed synthesis of furans from diazo malonate.

Wang and co-workers reported a copper-catalyzed cascade coupling/cyclization of terminal alkynes with α-alkyl-substituted diazoesters in 2011 [16]. 2,3,5-Trisubstituted furan derivatives were prepared with good efficiency and selectivity (Scheme 3.11). In their mechanistic study, they found that cyclopropenyl ester, which is a commonly proposed intermediate in rhodium-catalyzed transformation, is less likely to be the intermediate. They believed the reaction started with copper acetylide formation and then followed by the reaction of copper acetylide with diazo substrate leading to the formation of copper carbene species. Migratory insertion of alkynyl group to the carbenic carbon gives the key intermediates, which affords 3-alkynoate by the direct protonation (detected by-producte) or gives the final product through intramolecular nucleophilic attack of the carbonyl group to triple bond and followed by a subsequent proton transfer.

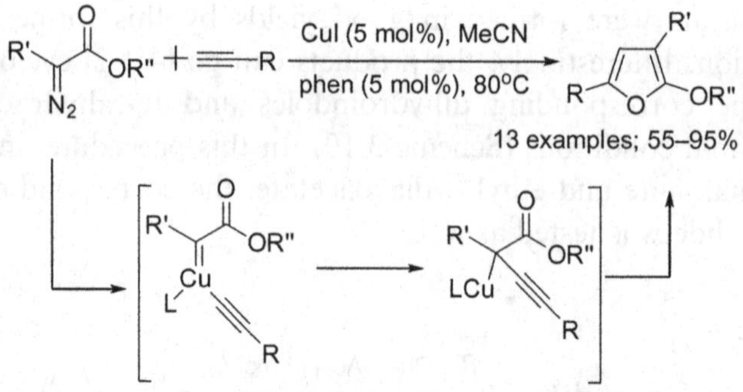

Scheme 3.11 Cu-catalyzed synthesis of furans from diazoesters.

Later on, Zhang and co-workers reported a cobalt-catalyzed version of this transformation [17]. This newly developed Co(II) metalloradical-catalyzed system offers an regioselective system for the synthesis of furans with alkynes and diazocarbonyls as the starting materials (Scheme 3.12). This metalloradical cyclization system has a wide substrate scope and good functional group tolerance; multisubstituted furans with diverse functionalities were formed in good yields. From a mechanistic point of view, the reaction starts between the reaction of Co(II)-porphyrin complexes and α-diazocarbonyls to generate Co(III)-carbene radicals and then undergo a new type of tandem radical addition reaction with alkynes that affords the final five-membered furans.

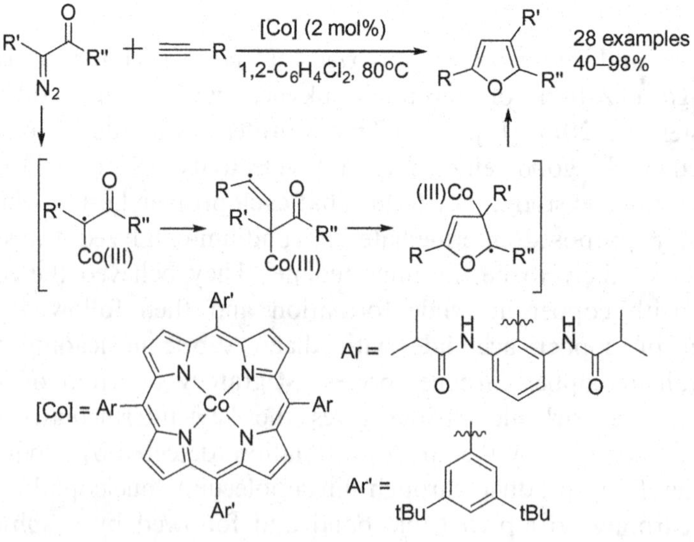

Scheme 3.12 Co-catalyzed synthesis of furans from diazoesters.

More recently, Lee and co-workers developed a Ru-catalyzed procedure for the synthesis of highly functionalized furans with readily available cyclic and acyclic diazodicarbonyl compounds and terminal alkynes as the substrates [18]. A variety of diverse furan derivatives were prepared in good yields by this procedure (Scheme 3.13).

Scheme 3.13

Tris(triphenylphosphine)ruthenium(II)   dichloride   ([Ru(PPh$_3$)$_3$Cl$_2$]; 0.02 mol, 2 mol%) was added to a solution of a cyclic diazodicarbonyl compound (1.0 mmol, 1 equiv.) and terminal alkyne (3.0 mmol, 3 equiv.) in toluene (2.0 mL) at room temperature. The reaction mixture was stirred at 70°C for the indicated time and then cooled to room temperature. Water (15 mL) was added and the solution was extracted with ethyl acetate (EA; 3 × 15 mL). Evaporation of the solvent and purification by column chromatography on silica gel using hexane/ethyl acetate (6:1) as eluent gave the product.

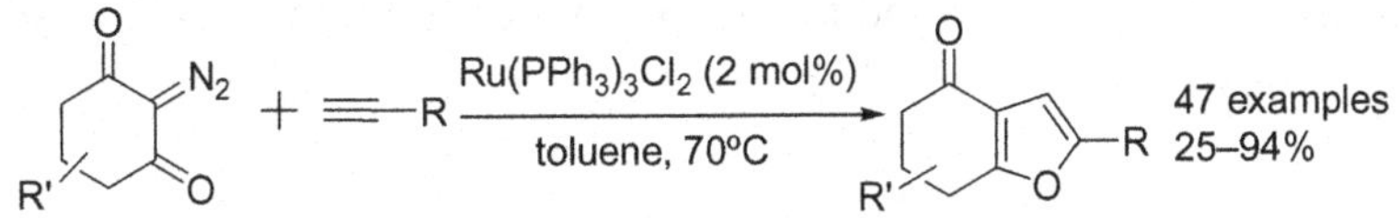

Scheme 3.13 Ru-catalyzed synthesis of furans from diazodicarbonyl compounds.

As ylides and carbenes (or carbenoids) are related species, a procedure with iodonium ylide as the substrate was developed in 1997 [19]. The iodonium ylide was prepared in 93% yield from the reaction of β-ketosulfone with phenyl iodosyl bis(trifluoroacetate). Moderate yield of the desired furan derivative can be obtained with alkyne and cyclopropanes can be formed with various alkenes (Scheme 3.14).

Later on, Hadjiarapoglou and co-workers reported the cyclization of 2-phenyliodonio-5,5-dimethyl-1,3-dioxacyclohexanemethylide with acetylenes and nitriles [20]. With Rh$_2$(OAc)$_4$ as the catalyst, the corresponding furans and oxazoles were formed in moderate yields. The possibilities of applying carbomethoxy iodonium ylides, generated from methyl acetoacetate and methyl malonate, were exploited by the same group [21]. Cyclopropenes and various heterocyclic" compounds can be produced in good yields. Lee and Yoon further studied this transformation with electron deficient and conjugated alkynes [22]. The corresponding furan derivatives were isolated in moderate yields.

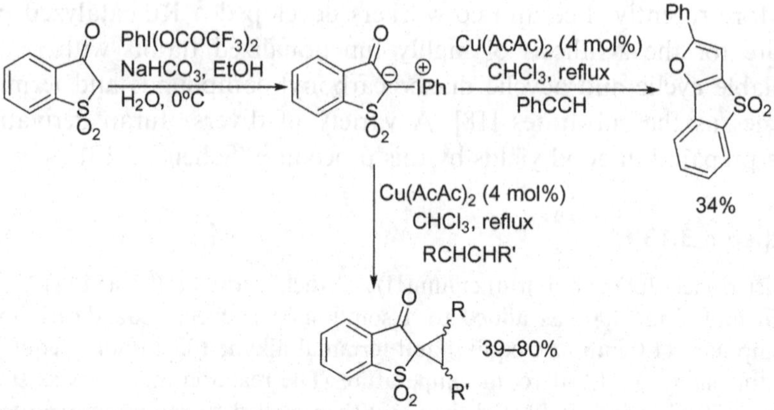

Scheme 3.14 Cu-catalyzed synthesis of furans from ylide.

Yan et al. developed a straightforward method for the construction of polysubstituted furans with alkynoates and 1,3-dicarbonyl compounds as the substrates in 2010 [23]. Oxygen was used as the oxidant in this procedure and moderate to good yields of the desired furans can be produced (Scheme 3.15).

Scheme 3.15

An oven-dried Schlenk tube was charged with CuI (9.5 mg, 0.05 mmol), 1,3-dicarbonyl compound (0.50 mmol), and alkyne (0.50 mmol). The Schlenk tube was sealed and then evacuated and backfilled with oxygen (3 cycles). Then DMF (2 mL) was added to the reaction system. The reaction was stirred at 110°C under O_2 (1 atm) for 4 h. After cooling to r.t., the solvent diluted with Et_2O (10 mL), washed with brine (5 mL) and then dried over anhyd Na_2SO_4. After the solvent was evaporated in vacuo, the residues were purified by column chromatography, eluting with PE-EtOAc (10:1) to afford pure product.

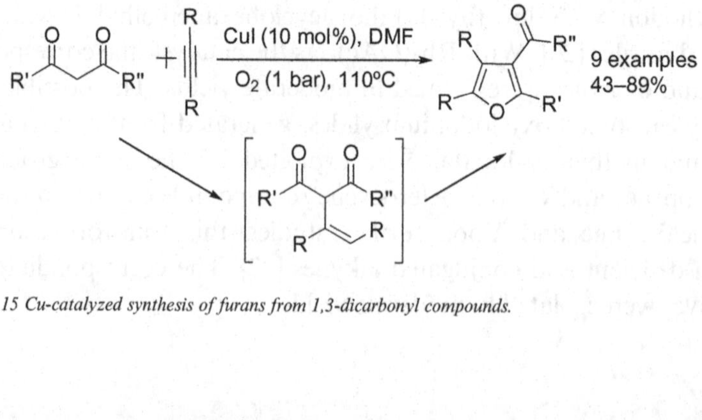

Scheme 3.15 Cu-catalyzed synthesis of furans from 1,3-dicarbonyl compounds.

In 2011, Deepthi and co-workers reported a cerium(IV) ammonium nitrate mediated version of this transformation [24]. With over excess of ceric ammonium nitrate (CAN) as the oxidation and radical initiator, 1,3-dicarbonyl compounds added to terminal acetylenes and yielded the desired multisubstituted furan derivatives in moderate to good yields (Scheme 3.16).

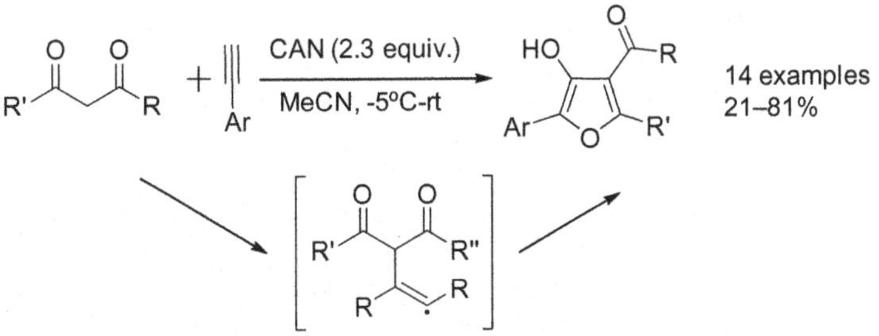

Scheme 3.16 CAN-promoted synthesis of furans from 1,3-dicarbonyl compounds.

In 2014, a gold-catalyzed oxidative cross-coupling of 1,3-dicarbonyl compounds with terminal alkynes was developed by You and co-workers [25]. 3-Alkynyl polysubstituted furans were prepared in good yields under mild reaction conditions with complete regio-control and wide substrate scope and high functional-group tolerance (Scheme 3.17). Hypervalent iodine reagent was applied as the oxidant here.

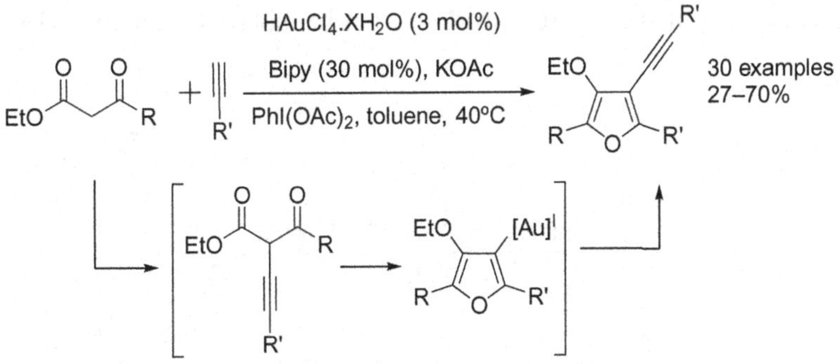

Scheme 3.17 Au-catalyzed synthesis of furans from 1,3-dicarbonyl compounds.

In 2009, Jiang and co-workers developed a novel method for the direct construction of polysubstituted furans [26]. This transformation involves Sn(II)- and Cu(I)-involved addition/oxidative cyclization of alkynoates and 1,3-dicarbonyl compounds in the presence of 2,3-dichloro-5,6-dicyanobenzoquinone (DDQ). Later on, they reported a two-step procedure for the synthesis of the same class of furans from similar substrates [27]. In this new methodology, air was used as the oxidant and CuBr as the catalyst. AcOH was applied as acid to replace $SnCl_2$. Moderate to good yields of the products can be isolated (Scheme 3.18).

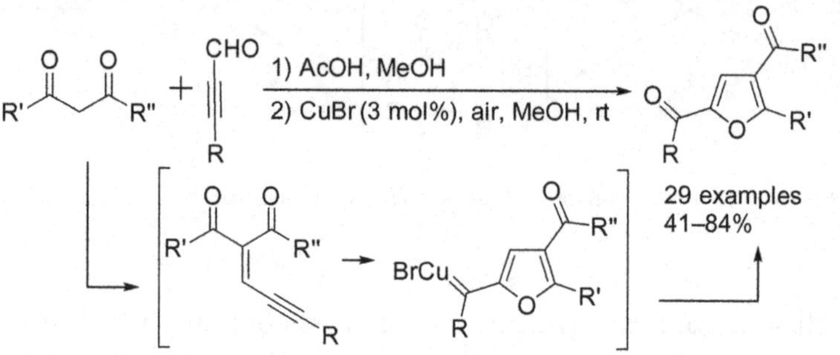

Scheme 3.18 Cu-catalyzed synthesis of trisubstituted furans from 1,3-dicarbonyl compounds.

Zhang and Xiao reported a catalyst-controlled regioselective synthesis of furans and 4H-pyrans from 2-(1-alkynyl)-2-alken-1-ones and 1,3-dicarbonyl compounds [28]. From the same substrates, furans can be isolated when palladium was applied as the catalyst while 1,8-Diazabicyclo[5.4.0]undec-7-ene (DBU) give 4H-pyrans as the main product via Michael addition and cyclization reaction sequence. Moderate to excellent yields of the desired products can be isolated (Scheme 3.19).

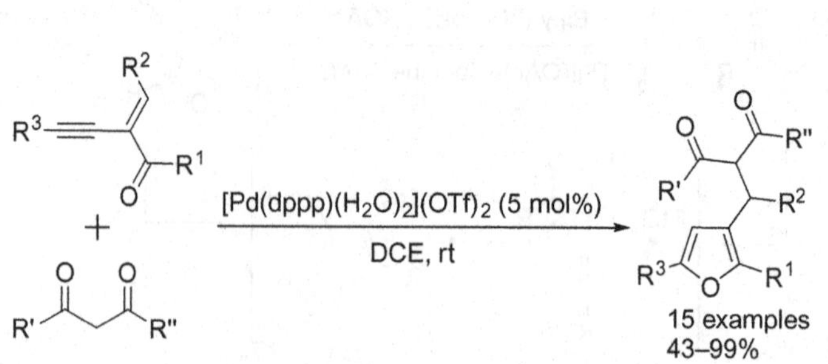

Scheme 3.19 Pd-catalyzed synthesis of furans from 1,3-dicarbonyl compounds.

In 1997, Lee and co-workers reported the synthesis of 3-acylfurans by using silver(I)/Celite as the promoter [29]. Started from dicarbonyl compounds and vinyl sulfides, the desired products were formed, followed by cycloaddition, NaIO$_4$ oxidation, and *syn*-elimination sequences. Later on, they found this procedure can be performed in a one-pot one-step manner in the synthesis of medium- and large-sized ring substituted furans from 1,3-dicarbonyl compounds with vinyl sulfides [30]. Moderate to good yields of the desired furans can be formed (Scheme 3.20).

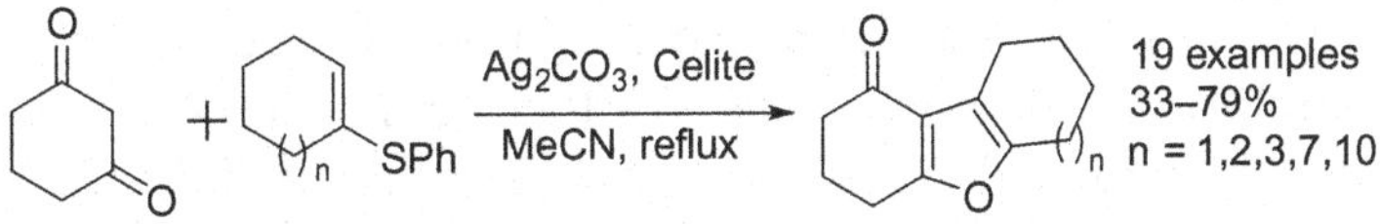

Scheme 3.20 Ag-mediated synthesis of furans from 1,3-dicarbonyl compounds.

A one-pot procedure for the synthesis of substituted furans was developed in 2007 [31]. Good yields can be achieved by using but-2-ene-1,4-diones and acetoacetates as the substrates and InCl$_3$ as the catalyst (Scheme 3.21). Various Lewis acids were examined; moderate yields can be achieved with FeCl$_3$ or ZnCl$_2$ as the catalyst.

Scheme 3.21

To a solution of but-2-ene-1,4-dione (264 mg, 1 mmol) and methyl acet-oacetate (128 mg, 1.1 mmol) in dry i-PrOH (7 ml) anhydrous InCl$_3$ (45 mg, 20 mol %) was added. The reaction mixture was then stirred under reflux at 80–90°C for 5.5 h. After complete disappearance of the starting material [monitored by TLC using petroleum ether-chloroform (6:4)], the solvent was removed from the reaction mixture on a rotary evaporator. The residue was then diluted with water (10 ml) and extracted with CHCl$_3$ (3 × 25 ml). The organic layer was separated, washed with brine and then dried over anhydrous Na$_2$SO$_4$. Removal of the solvent resulted in a solid which was chromatographed over silica gel using petroleum ether and an increasing proportion of ethyl acetate as eluent. Petroleum ether–ethyl acetate (96:4) eluent gave a solid which was recrystallized from chloroform-petroleum ether (2:8) to give the pure product.

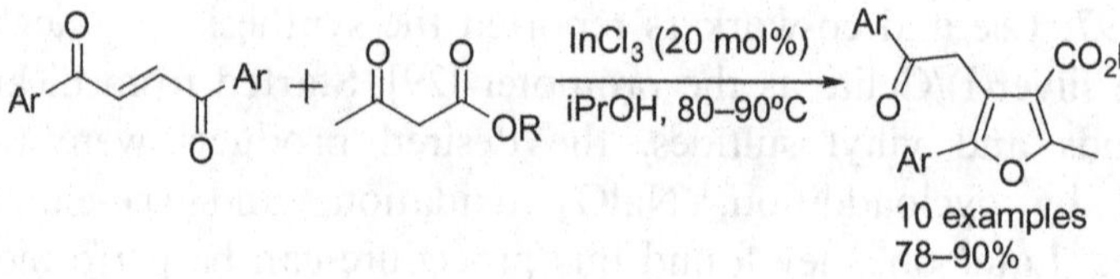

Scheme 3.21 In-catalyzed synthesis of furans from acetoacetate.

Recently, a methodology for the preparation of tetrasubstituted 3-acylfuran derivatives was reported by Schickmous and Christoffers [32]. By condensing of acetylacetone with alicyclic and heterocyclic α-hydroxy-β-oxo esters under acidic conditions [0.1 equiv. cerium(III) salt in AcOH], moderate yields of the desired products can be obtained. [b]-Annulated 4-acyl-3-hydroxy-5-methylfuran-2-carboxylates were found to be the common intermediates, and then the reaction proceeded along different pathways depending on the starting materials used. Alicyclic α-hydroxy-β-oxo esters gave cycloalkane-annulated furans as products. The ester group was lost after saponification and decarboxylation at elevated temperature. A thiopyrano [3,4−b]-annulated 2H-furan was obtained when starting from a tetrahydrothiopyranone derivative. With piperidones or a tetrahydropyranone as starting materials, the heterocyclic six-membered ring was cleaved in a *retro*-Mannich type reaction and one carbon atom was lost. 4-Acetyl-5-methylfuran-2-carboxylates with a 2-amino or 2-hydroxyethyl substituent in the 3-position are obtained in these cases.

An anhydrous iron(III) chloride-catalyzed synthesis of Garcia Gonzalez polyhydroxyalkyl- and *C*-glycosylfurans from unprotected sugar aldoses with β-keto esters by Knoevenagel condensation was developed in 2009 [33]. Good yields can be achieved easily under mild conditions (Scheme 3.22).

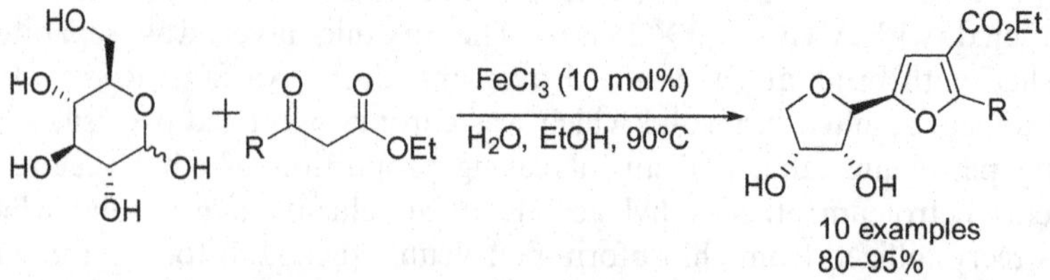

Scheme 3.22 Fe-catalyzed synthesis of furans from sugar aldoses.

Hajra and co-workers developed a copper-mediated intermolecular annulation of alkyl ketones and β-nitrostyrenes for the regioselective synthesis of multisubstituted furan derivatives in 2015 [34]. Both cyclic and acyclic ketones are applicable and the desired furans were formed in good yields via radical intermediates (Scheme 3.23).

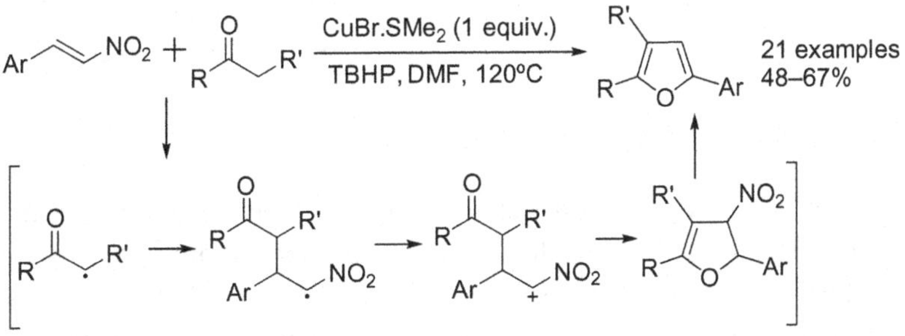

Scheme 3.23 Cu-mediated synthesis of furans from β-nitrostyrenes.

Wang's group developed a methodology for furans preparation based on the homo-cyclization of 2-aryl acetophenones [35]. Tetrasubstituted furans are formed in good yields with the corresponding 2,3-disubstituted-1,4-dicarbonyl compounds as the key intermediates (Scheme 3.24). Oxygen was applied as the terminal oxidant here and it showed high functional group tolerance. From mechanistic study, the reaction was proven go through a single-electron transfer (SET) process.

Scheme 3.24

Substrate (1.0 mmol), Ag$_2$O (23.2 mg, 0.1 mmol), and Cu(OAc)$_2 \cdot$H$_2$O (39.8 mg, 0.2 mmol) were added to a round-bottom flask (10 mL). Xylene (1.5 mL) and TFA (0.12 mL, 1.6 mmol) were slowly added to the mixture. The flask was then evacuated and backfilled with O$_2$, and the reaction mixture was stirred at 140°C. The progress of the reaction was monitored by TLC analysis. Upon completion, the crude mixture was cooled to room temperature and filtered through a small pad of silica gel. The filtrate was concentrated in vacuo, and the residue was purified by chromatography on a silica gel column to afford the pure product.

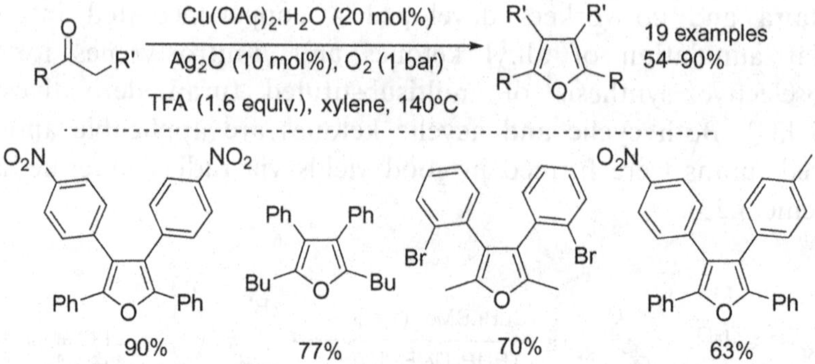

Scheme 3.24 Cu-catalyzed synthesis of furans from ketones.

Ishii and co-workers developed a new synthetic method for transforming acrylates and aldehydes to the corresponding substituted furans [36]. By using Pd(OAc)$_2$ as the catalyst and combined with molybdovanadophosphoric acid and Lewis acid under atmospheric dioxygen, good yields of the desired furans can be observed (Scheme 3.25). From mechanistic point of view, the reaction started with palladium-catalyzed acetalization of acrylates with methanol and followed by the reaction of the resulting acetals with aldehydes. Here, the Pd-(II)-catalyzed acetalization of acrylate with methanol to give acetal is a key step. The aldol-type condensation of the resulting acetal with aldehyde by CeCl$_3$ affords an α,β-unsaturated carbonyl condensate on which subsequent enolization by Pd(II) followed by intramolecular cyclization produces methyl furoate after Pd-H elimination from a dihydrofuran intermediate. Finally, the reduced Pd(0) was reoxidized to Pd(II) by the action of the HPMo11V/O$_2$ reoxidation system.

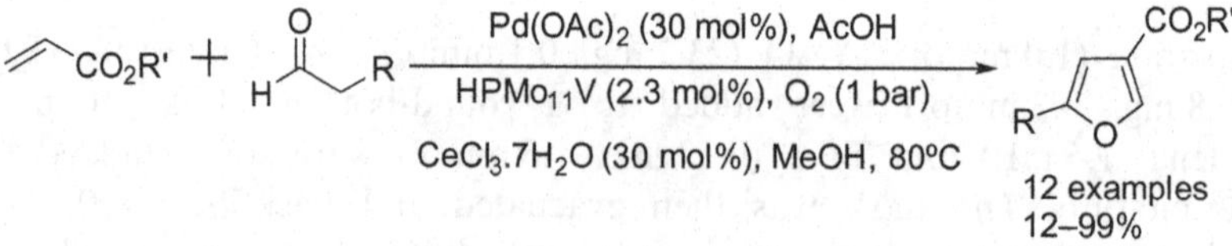

Scheme 3.25 Pd-catalyzed synthesis of furans from acrylates.

Jiang and co-workers developed an interesting procedure for the synthesis of 2,5-disubstituted 3-iodofurans in 2011 [37]. The procedure is based on palladium-catalyzed Sonogashira coupling of (Z)-β-bromoenol acetates with terminal alkynes, and then followed by intramolecular iodocyclization. Good to excellent yields of the furans were formed (Scheme 3.26). In this study, the authors found that the iodocyclization was sensitive to the nature of the solvent and the structure of conjugate enyne acetates. Notably, the 2,5-disubstituted 3-iodofurans are attractive for the prearation of more highly substituted furans.

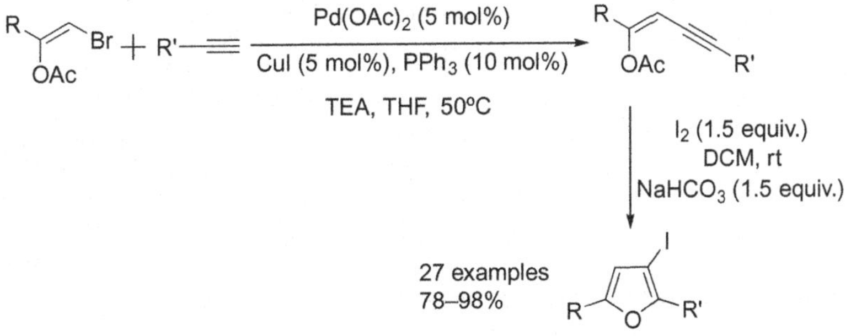

Scheme 3.26 Pd-catalyzed synthesis of furans from enyne acetates.

The reaction between cuprous acetylides with α-halo ketones and the corresponding furans was reported in as early as in 1967 [38]. By heating the two reagents together at 140–240°C for 5 min, low to moderate yields of the desired furans can be obtained.

Balme and co-workers developed a procedure for the synthesis of 2-substituted furo[3,2−c]pyridin-4-ones in 2008 [39]. They show that 3-iodopyridin-2-ones can be transformed into the desired 2-substituted furan derivatives in good yields through *in situ* sequential Sonogashira-acetylide coupling, dealkylation, and furan annulation reactions (Scheme 3.27a). For the dealkylation process, an Et₃N-induced S_N2 mechanism was proven. Additionally, structurally related compounds like pyrone and coumarin derivatives can be applied as substrates as well. They developed a procedure for the substituted furo[2,3−b]pyridines with 3-iodo-2-pyridones, terminal alkynes, and aryl iodides as the substrates as well (Scheme 3.27b) [40].

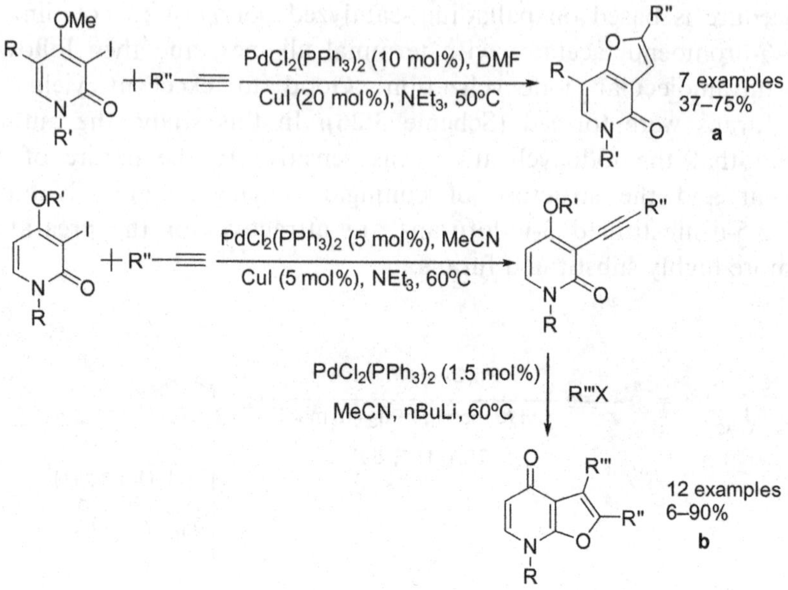

Scheme 3.27 Pd-catalyzed synthesis of furopyridines from 3-iodo-2-pyridones.

Liang and co-workers reported a three-component cyclization-coupling reaction to prepare furans in 2005 [41]. With palladium as the catalyst, polysubstituted furans were formed in good yields from readily available substrates (Scheme 3.28). The authors proposed that the reaction started with the oxidative addition of the organic halide to the Pd⁰ catalyst generated a δ-arylPd^{II}X complex. Then the propargyl carbonate reacts with this palladium complex to give the π-palladium complex which further transformed into the corresponding arylalkynylpalladium. After reacted with methyl acetoacetate, furans can be formed after intramolecular cyclization.

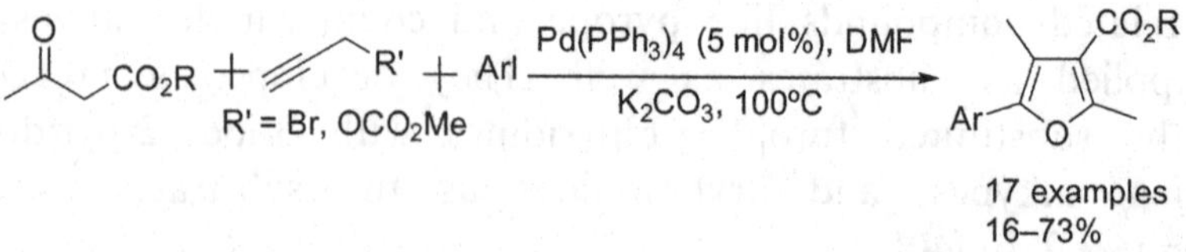

Scheme 3.28 Pd-catalyzed synthesis of furans from aryl halides.

A Pd/Cu-catalyzed cascade Sonogashira coupling/cyclization reactions for the synthesis of highly substituted 3-formyl furans was developed in 2011 [42]. By using α-bromoenaminones and terminal alkynes as the substrates, good yields of the desired furans can be produced by this sequential reaction (Scheme 3.29). For the intermediate, alkynylated enaminones were found to be the one.

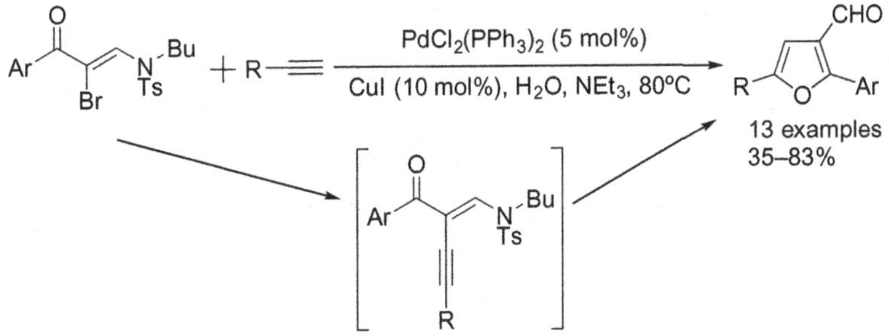

Scheme 3.29 Pd-catalyzed synthesis of furans from α-bromoenaminones.

Müller and co-workers developed a novel consecutive three-component coupling-addition-cyclocondensation reaction for the synthesis of 3-halofurans and 3-chloro-4-iodofurans [43]. With acyl chlorides and alkynes as the substrates, valuable 3-halofurans and 3-chloro-4-iodofurans can be produced under the assistant of acid (Scheme 3.30a). This type of transformation was reported with $ZnBr_2$ as the catalyst as well [44]. α,β-Acetylenic ketones were synthesized from the reaction of acid chlorides and acetylenic compounds in the presence of $ZnBr_2$ and DIEA in acetonitrile. In the case of the acetylenic ketones having nearby methylene unit, 2,5-disubstituted furan derivatives could be synthesized under the same reaction conditions (Scheme 3.30b).

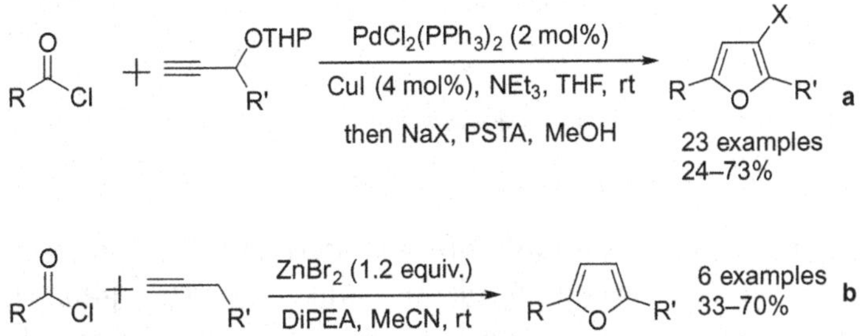

Scheme 3.30 Synthesis of furans from acid chloride.

The cyclization reaction of internal alkynes to give furans was achieved as well [45]. A number of heterocycles were synthesized regioselectively in good yields by applying appropriate vinylic halides and internal alkynes as substrates and palladium as the catalyst (Scheme 3.31). Rhodium catalyst was applied in furan synthesis as well. By applying terminal alkynes and ethyl diazoacetate as the substrates, furans can be produced in good yields with cyclopropenes as the intermediates [46].

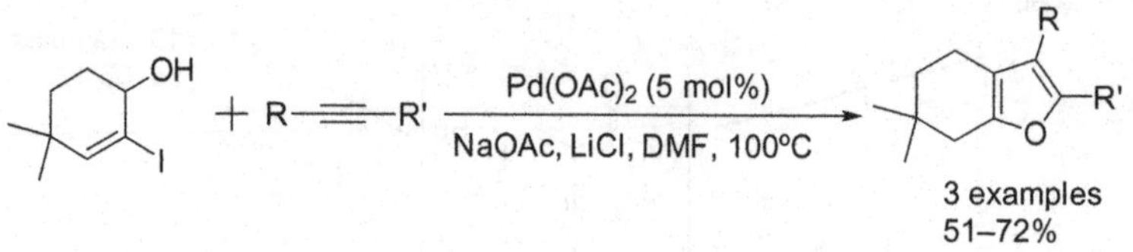

Scheme 3.31 Pd-catalyzed synthesis of furans from vinylic halides.

A silver-catalyzed furan synthesis was developed in 2013 by Jiang and co-workers [47]. With ethyl benzoylacetates and arylethynyl bromide as the starting materials and $AgNO_3$ as the catalyst, 2,3,4-trisubstituted furans were produced in a one-pot manner in good yields (Scheme 3.32). Regarding the mechanism, vinyl bromide intermediate was generated via an intermolecular nucleophilic addition of ethyl 2-pyridylacetate to phenylethynyl bromide with 1,4-Diazabicyclo[2.2.2] octane (DABCO) and $AgNO_3$ as the promotor. Subsequently, the vinyl bromide intermediate undergoes the enol-ketone equilibrium and finally gives the final product after the loss of HBr.

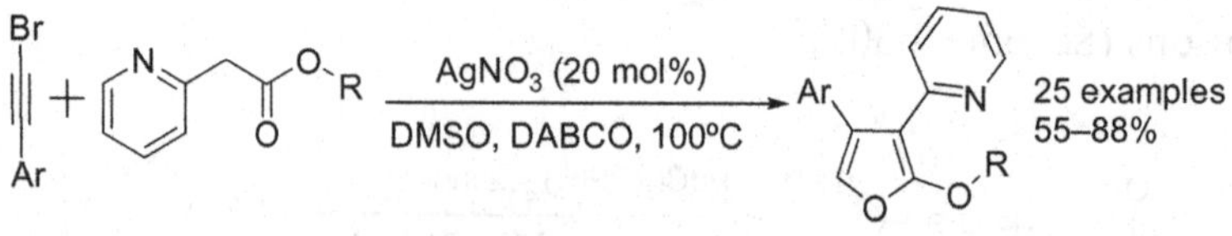

Scheme 3.32 Ag-catalyzed synthesis of furans from arylethynyl bromide.

Copper-mediated 2,3-dihydrofuran derivatives synthesis from 2,2-dibromo 1,3-diones and olefins was reported [48]. In this procedure, furans can be formed when alkynes were applied as the coupling partner.

Beifuss and co-workers developed a copper-catalyzed domino reaction for the synthesis of furans from 2,3-dihalo-1-propenes and β-ketoesters or 1,3-diketones [49]. The reactions were carried out in dimethylformamide (DMF) at 120°C with Cs_2CO_3 as a base and hydroquinone as an additive. Various 2,3,5-trisubstituted furans and related compounds were produced with yields up to 96% (Scheme 3.33). The highly regioselective domino process is based on an intermolecular *C*-allylation followed by an intramolecular Ullmann-type *O*-vinylation and a double bond isomerization.

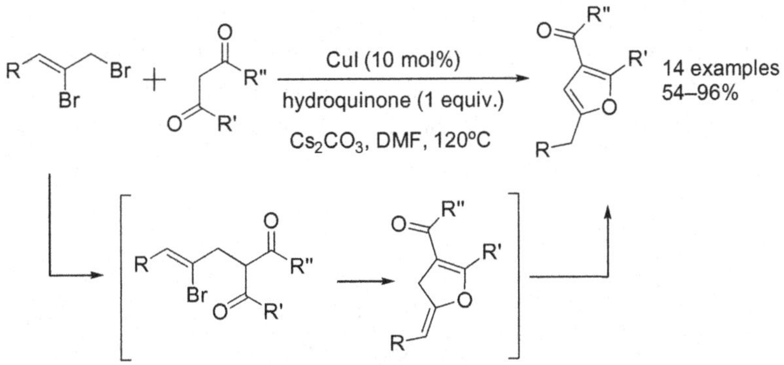

Scheme 3.33 Cu-catalyzed synthesis of furans from 2,3-dihalo-1-propenes.

A palladium-catalyzed oxidative difunctionalization of enol ethers with 1,3-dicarbonyl compounds to construct trisubstituted furans was reported by Jiang and co-workers in 2013 [50]. Oxygen was used as the sole oxidant in this procedure for regenerating the Pd(II) catalyst. Numbers of furans were formed in good to excellent yields with carboetherification of enol ethers with 1,3-carbonyls as the key step (Scheme 3.34).

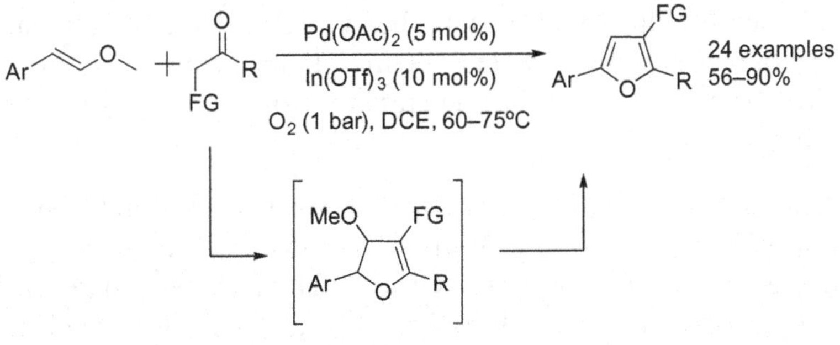

Scheme 3.34 Pd-catalyzed synthesis of furans from enol ethers.

Liu and co-workers demonstrated an efficient multicomponent coupling reaction of phenylglyoxal derivatives, secondary amines, and terminal alkynes in 2013 [51]. This cascade transformation using $AuBr_3$ as the catalyst under N_2 atmosphere in methanol, good yields of the furans can be obtained after coupling/cycloisomerization reactions (Scheme 3.35). This methodology provides a new strategy for the synthesis of 3-substituted furans with high atom economy and high catalytic efficiency. In detail, reaction started with a gold-catalyzed three-component coupling of a phenylglyoxal, an amine, and an alkyne to afford NR^1R^2-substituted propargylic intermediate via a Mannich–Grignard reaction. Then coordination of the triple bond in propargylic intermediate to the gold catalyst enhances the electrophilicity of the alkyne, and the subsequent nucleophilic attack of the oxygen lone pair would produce the cation intermediate, which undergoes deprotonation followed by demetalation to afford the final product.

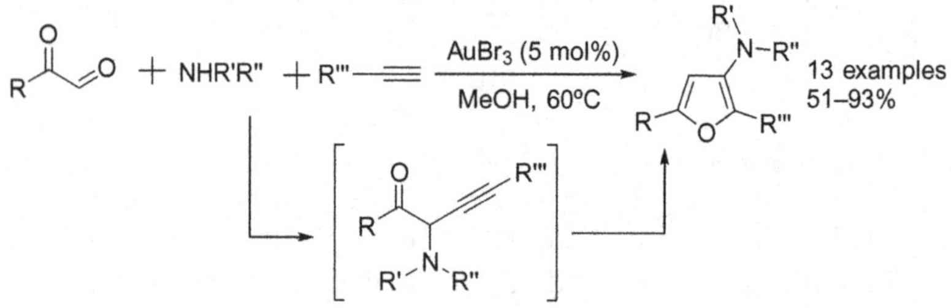

Scheme 3.35 Au-catalyzed synthesis of furans from phenylglyoxal derivatives.

Tanaka's group demonstrated a cationic rhodium(I)/Segphos or H^8-BINAP complex-catalyzed enantioselective and diastereoselective cotrimerization of commercially available monoenes and dialkyl acetylenedicarboxylates to the corresponding functionalized furylcyclopropanes in 2008 [52]. Through mechanistic studies, the furylcyclopropanes might be generated through rhodacyclopentene intermediates.

Olefin cross-metathesis (CM) reaction was explored and applied in furan synthesis as well [53]. With allylic alcohols and methyl vinyl ketones as the substrates, and the Grubbs–Hoveyda second-generation catalyst (G-H II) as the catalyst, 2,5-di- or 2,3,5-tri-substituted furans were produced in good yields.

REFERENCES

[1] a. J. Tsuji, H. Watanabe, I. Minami, I. Shimizu, J. Am. Chem. Soc. 107 (1985) 2196–2198.
    b. I. Minami, M. Yuhara, H. Watanabe, J. Tsuji, J. Organomet. Chem. 334 (1987) 225–242.
    c. N. Greeves, J.S. Torode, Synthesis (1993) 1109–1112.

[2] B.M. Trost, M.C. McIntosh, J. Am. Chem. Soc. 117 (1995) 7255–7256.

[3] a. V. Cadierno, J. Gimeno, N. Nebra, Adv. Synth. Catal. 349 (2007) 382–394.
    b. V. Cadierno, J. Díez, J. Gimeno, N. Nebra, J. Org. Chem. 73 (2008) 5852–5858.

[4] a. W. Ji, Y. Pan, S. Zhao, Z. Zhan, Synlett (2008) 3046–3052.
    b. Z. Zhan, X. Cai, S. Wang, J. Yu, H. Liu, Y. Cui, J. Org. Chem. 72 (2007) 9838–9841.

[5] a. Y. Pan, S. Zhao, W. Ji, Z. Zhan, J. Comb. Chem. 11 (2009) 103–109.
    b. Z. Zhan, S. Wang, X. Cai, H. Liu, J. Yu, Y. Cui, Adv. Synth. Catal. 349 (2007) 2097–2102.

[6] Y. Nishibayashi, M. Yoshikawa, Y. Inada, M.D. Milton, M. Hidai, S. Uemura, Angew. Chem. Int. Ed. 42 (2003) 2681–2684.

[7] P. Lenden, D.A. Entwistle, M.C. Willis, Angew. Chem. Int. Ed. 50 (2011) 10657–10660.

[8] D. Nitsch, T. Bach, J. Org. Chem. 79 (2014) 6372–6379.

[9] N. Monteiro, G. Balme, J. Org. Chem. 65 (2000) 3223–3226.

[10] M. Nakamura, M. Yamane, H. Sakurai, K. Narasaka, Heterocycles 59 (2003) 333–345.

[11] a. H. Cao, H. Jiang, W. Yao, X. Liu, Org. Lett. 11 (2009) 1931–1933.
     b. H. Cao, H. Jiang, H. Huang, Synthesis (2011) 1019–1036.

[12] I. Minami, M. Yuhara, J. Tsuji, Tetrahedron Lett. 28 (1987) 629–632.

[13] a. M.C. Pirrung, J. Zhang, A.T. Morehead Jr., Tetrahedron Lett. 35 (1994) 6229–6230.
     b. M.C. Pirrung, F. Blume, J. Org. Chem. 64 (1999) 3642–3649.

[14] a. Y.R. Lee, J.Y. Suk, B.S. Kim, Tetrahedron Lett. 40 (1999) 6603–6607.
     b. Y.R. Lee, J.Y. Suk, Tetrahedron Lett. 41 (2000) 4795–4799.

[15] H. Li, R.P. Hsung, Org. Lett. 11 (2009) 4462–4465.

[16] L. Zhou, J. Ma, Y. Zhang, J. Wang, Tetrahedron Lett. 52 (2011) 5484–5487.

[17] X. Cui, X. Xu, L. Wojtas, M.M. Kim, X.P. Zhang, J. Am. Chem. Soc. 134 (2012) 19981–19984.

[18] L. Xia, Y.R. Lee, Eur. J. Org. Chem. (2014) 3430–3442.

[19] L.P. Hadjiarapoglou, K. Schank, Tetrahedron 53 (1997) 9365–9376.

[20] E.P. Gogonas, L.P. Hadjiarapoglou, Tetrahedron Lett. 41 (2000) 9299–9303.

[21] C. Batsila, G. Kostakis, L.P. Hadjiarapoglou, Tetrahedron Lett. 43 (2002) 5997–6000.

[22] Y.R. Lee, S.H. Yoon, Synth. Commun. 36 (2006) 1941–1951.

[23] R. Yan, J. Huang, J. Luo, P. Wen, G. Huang, Y. Liang, Synlett (2010) 1071–1074.

[24] A. Sivan, A. Deepthi, V. Nandialath, Synthesis (2011) 2466–2470.

[25] Y. Ma, S. Zhang, S. Yang, F. Song, J. You, Angew. Chem. Int. Ed. 53 (2014) 7870–7874.

[26] W. Liu, H. Jiang, M. Zhang, C. Qi, J. Org. Chem. 75 (2010) 966–968.

[27] H. Cao, H. Zhan, J. Cen, J. Lin, Y. Lin, Q. Zhu, et al., Org. Lett. 15 (2013) 1080–1083.

[28] Y. Xiao, J. Zhang, Chem. Commun. (2009) 3594–3596.

[29] Y.R. Lee, N.S. Kim, B.S. Kim, Tetrahedron Lett. 38 (1997) 5671–5674.

[30] Y.R. Lee, J.Y. Suk, B.S. Kim, Org. Lett. 2 (2000) 1387–1389.

[31] S. Dey, D. Nandi, P.K. Pradhan, V.S. Giri, P. Jaisankar, Tetrahedron Lett. 48 (2007) 2573–2575.

[32] B. Schickmous, J. Christoffers, Eur. J. Org. Chem. (2014) 4410–4416.

[33] L. Nagarapu, M.V. Chary, A. Satyender, B. Supriya, R. Bantu, Synthesis (2009) 2278–2282.

[34] M. Ghosh, S. Mishra, A. Hajra, J. Org. Chem. 80 (2015) 5364–5368.

[35] S. Mao, Y.R. Gao, S.L. Zhang, D.D. Guo, Y.Q. Wang, Eur. J. Org. Chem. (2015) 876–885.

[36] K. Tamaso, Y. Hatamoto, Y. Obora, S. Sakaguchi, Y. Ishii, J. Org. Chem. 72 (2007) 8820–8823.

[37] Z. Chen, G. Huang, H. Jiang, H. Huang, X. Pan, J. Org. Chem. 76 (2011) 1134–1139.

[38] K. Gump, S.W. Moje, C.E. Castro, J. Am. Chem. Soc. 89 (1967) 6670–6671.

[39] D. Conreaux, S. Belot, P. Desbordes, N. Monteiro, G. Balme, J. Org. Chem. 73 (2008) 8619–8622.

[40] E. Bossharth, P. Desbordes, N. Monteiro, G. Balme, Org. Lett. 5 (2003) 2441–2444.

[41] X. Duan, X. Liu, L. Guo, M. Liao, W. Liu, Y. Liang, J. Org. Chem. 70 (2005) 6980–6983.

[42] J. Yang, C. Wang, X. Xie, H. Li, E. Li, Y. Li, Org. Biomol. Chem. 9 (2011) 1342–1346.

[43] a. A.S. Karpov, E. Merkul, T. Oeser, T.J.J. Müller, Eur. J. Org. Chem. (2006) 2991–3000.
     b. A.S. Karpov, E. Merkul, T. Oeser, T.J.J. Müller, Chem. Commun. (2005) 2581–2583.

[44] K.Y. Lee, M.J. Lee, J.N. Kim, Tetrahedron 61 (2005) 8705–8710.

[45] R.C. Larock, M.J. Doty, X. Han, Tetrahedron Lett. 39 (1998) 5143–5146.

[46] H.M.L. Davies, K.R. Romines, Tetrahedron 44 (1988) 3343–3348.

[47] W. Zeng, W. Wu, H. Jiang, Y. Sun, Z. Chen, Tetrahedron Lett. 54 (2013) 4605–4609.

[48] J. Yoshida, S. Yano, T. Ozawa, N. Kawabata, J. Org. Chem. 50 (1985) 3467–3473.

[49] D. Schmidt, C.C. Malakar, U. Beifuss, Org. Lett. 16 (2014) 4862–4865.

[50] M. Zheng, L. Huang, W. Wu, H. Jiang, Org. Lett. 15 (2013) 1838–1841.

[51] J. Li, L. Liu, D. Ding, J. Sun, Y. Ji, J. Dong, Org. Lett. 15 (2013) 2884–2887.

[52] Y. Shibata, K. Noguchi, M. Hirano, K. Tanaka, Org. Lett. 10 (2008) 2825–2828.

[53] T.J. Donohoe, J.F. Bower, PNAS 107 (2010) 3373–3376.

CHAPTER 4

Synthesized by [4 + 1] Cyclization Reactions

In 2006, Alper's group reported a novel method for the synthesis of 3-substituted furans by hydroformylation of substituted propargylic alcohols (Scheme 4.1a) [1]. In this methodology, using $[Rh(OAc)_2]_2$ and PPh_3 as the catalytic system 3-substituted furans can be produced in good yields under mild conditions. In 2009, a rhodium-catalyzed carbonylative addition of arylboronic acids to propargylic alcohols to prepare furans was reported by Dheur et al. [2]. The reaction goes through gamma-hydroxy enones as the intermediate which subsequently cyclized through a dehydration step to yield the corresponding furan analogs (Scheme 4.1b).

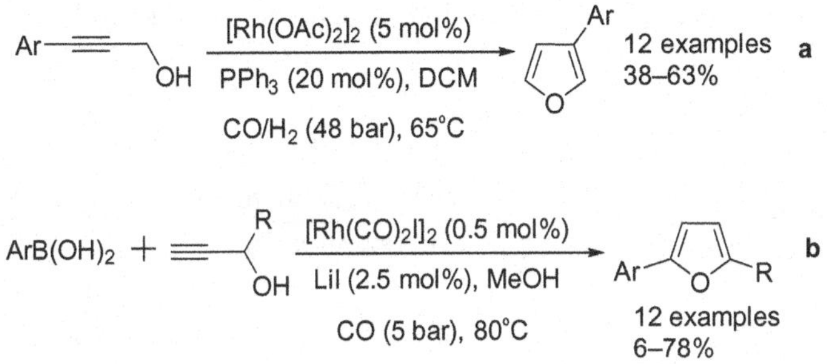

Scheme 4.1 Rh-catalyzed carbonylative synthesis of furans.

A manganese(III)-mediated reaction between 2-benzoyl-1, 4-naphthoquinones and 1,3-dicarbonyl compounds was described by Chuang and co-workers [3]. Naphtho[2,3−c]furan-4,9-diones and naphthacene-5,12-diones can be produced with moderate to high chemoselectivity, depending on the electronic effect of the benzoyl group substituent on the reactants. With ethyl benzoylacetate and 1,3-diketones, the novel naphtho[2,3−c]furan-4,9-diones were produced effectively with high selectivity. They developed procedures on silver (II)- and manganese(III)-mediated radical reactions of 2-substituted-1,4-naphthoquinones as well (Scheme 4.2) [4]. In this procedure, the acyl

Transition Metal Catalyzed Furan Synthesis. DOI: http://dx.doi.org/10.1016/B978-0-12-804034-8.00004-0

radicals were generated by the oxidative decarboxylation of α-keto acids with silver(I) nitrate and persulfate through radical addition to the $C = C$ bond of 2-(1-hydroxyalkyl)-1,4-naphthoquinones and 2-(1-amidoalkyl)-1,4-naphthoquinones. This reaction provides an effective method for the synthesis of naphtho[2,3−c]furan-4,9-diones and benzo[f]isoindole-4,9-diones. In the presence of O_2, manganese(III) acetate oxidation of β-keto esters also generates acyl radicals, which then undergo radical addition to 2-(1-amidoalkyl)-1,4-naphthoquinones, and subsequently, benzo[f]isoindole-4,9-diones were produced.

Scheme 4.2

A mixture of 2-(1-hydroxybenzyl)-1,4-naphthoquinone (0.50 mmol), 2-oxopropionic acid (1.51 mmol), silver(I) nitrate (0.20 mmol), and potassium persulfate (1.79 mmol) in acetonitrile (2 ml) and H_2O (6 ml) was heated at 70°C for 3 h. The reaction mixture was diluted with ethyl acetate (100 ml), washed with water (3 × 50 ml), dried (Na_2SO_4), and concentrated in vacuo. The crude product was purified by column chromatography over silica gel (20 g; ethyl acetate/hexane, 1:15) followed by recrystallization (ethyl acetate/hexane) to give the pure product.

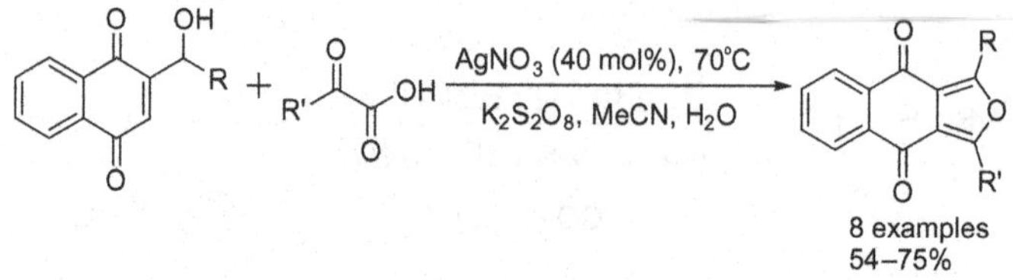

8 examples
54−75%

Scheme 4.2 Ag-mediated synthesis of furans from α-keto acids.

The use of 2,3-bis(trimethylsilyl)buta-1,3-diene and acyl chlorides as starting materials for furan synthesis was reported in 2007 [5]. With aluminium trichloride as the promotor, 2,5-disubstituted and, in many cases, 2,3,5-trisubstituted furans can be produced in good yields.

Cossy and co-workers developed a tandem Suzuki-Miyaura coupling/acid-catalyzed cyclization between vinyl ether boronates and vinyl halides to construct polysubstituted furans [6]. Starting from appropriately substituted 3-haloallylic alcohols and dihydrofuran-, dihydropyran-, or glycal-derived pinacol boronates, 2-(ω-hydroxyalkyl) furans were produced in good yields (Scheme 4.3).

Scheme 4.3

To a mixture of vinyl bromide or iodide (1 mmol, 1 equiv.), pinacol boronate (1.2 mmol, 1.2 equiv.), and $Pd(dppf)Cl_2 \cdot CH_2Cl_2$ catalyst (16 mg, 0.02 mmol, 2 mol%) in 1,4-dioxane (2 ml), 4 M aqueous NaOH (0.5 ml, 2 mmol, 2 equiv.) was added. The vial was flushed with argon and sealed. The reaction mixture was then heated under microwave irradiation at 100°C for 10 min. After cooling to rt, the complete consumption of the starting aryl halide was checked by TLC, the vial was opened, and $TsOH \cdot H_2O$ (418 mg, 2.2 mmol, 2.2 equiv.) was added in one portion. The resulting mixture was stirred at rt for 10 min (TLC control) and then poured into saturated aqueous $NaHCO_3$ (50 ml), extracted with EtOAc (3 × 30 ml), and the combined organic layers were washed with brine and dried over Na_2SO_4. The product was isolated by flash chromatography on silica gel.

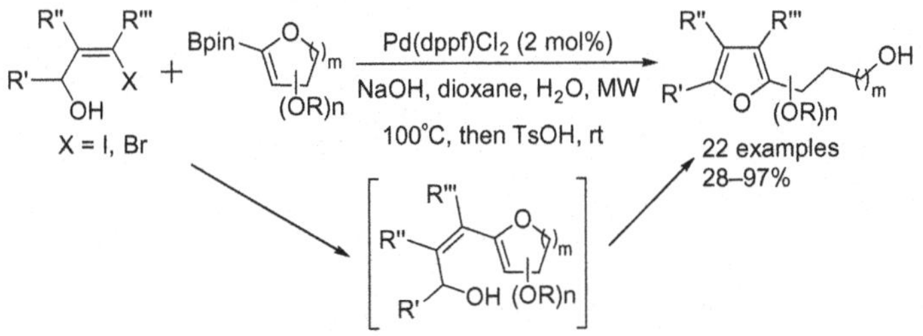

Scheme 4.3 Pd-catalyzed synthesis of 2-(ω-hydroxyalkyl)furans.

In 2010, Ma and co-workers developed a Cu^I/L-proline-catalyzed coupling reaction for the synthesis of furans [7]. With substituted 3-iodoprop-2-en-1-ols and 1-alkynes as the substrates, go through coupling and subsequent cyclization sequence, polysubstituted furans can be produced in good yields (Scheme 4.4).

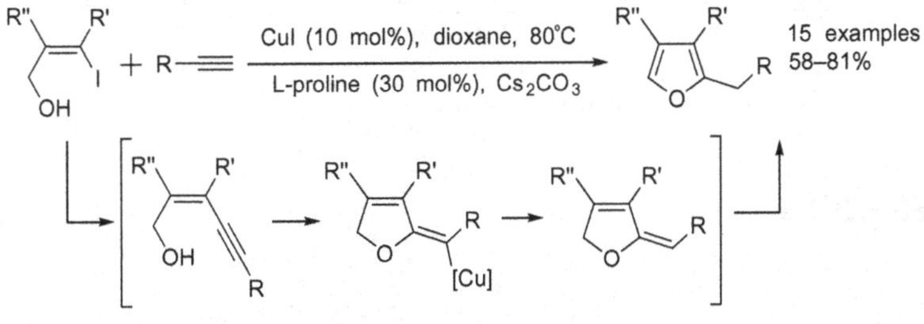

Scheme 4.4 Cu-catalyzed synthesis of furans from 3-iodoprop-2-en-1-ols.

A copper-catalyzed [4 + 1] cycloaddition reaction of α,β-acetylenic ketones with α-diazo esters was reported by Liang and co-workers in 2007 [8]. This methodology offers a direct route to highly substituted furans in good yields (Scheme 4.5). The possibility to involve cyclopropenyl ketone as the intermediate was considered.

Scheme 4.5

To a solution of α,β-acetylenic ketone (0.5 mmol) and CuI (0.1 mmol) in DCE (5 ml) heated at 90°C under argon was added a solution of a EDA (1.5 mmol) in DCE *via* syringe pump over 5 h. Next, the reaction was heated at 90°C with stirring for 5 h. The reaction was quenched with a saturated aqueous solution of ammonium chloride, and the mixture was extracted with dichloromethane (DCM). The combined organic extracts were washed with water and saturated brine. The organic layer was dried (Na$_2$SO$_4$) and concentrated in vacuo. The residue was purified by chromatography on silica gel to afford the pure product.

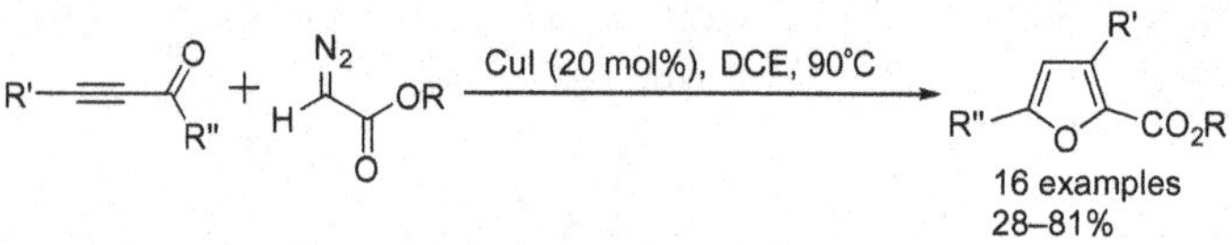

Scheme 4.5 Cu-catalyzed synthesis of furans from α-diazo esters.

Skrydstrup and co-workers reported an Au(I)-catalyzed hydroamination or hydration of 1,3-diynes to access 2,5-diamidopyrroles and 2,5-diamidofurans in 2010 [9]. Good yields of furans and pyrroles can be achieved under mild conditions (Scheme 4.6). Deuterated 2,5-disubstituted furans and 1,2,5-trisubstituted pyrroles can be prepared as well and also ^{18}O-labeled furans.

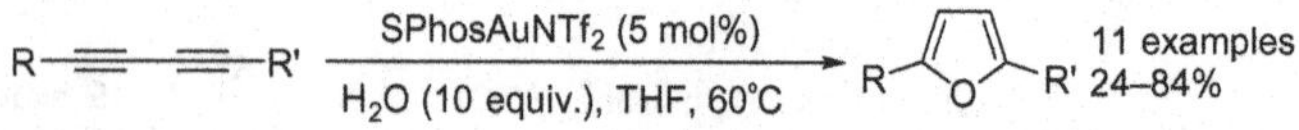

Scheme 4.6 Au-catalyzed synthesis of furans from 1,3-diynes.

REFERENCES

[1] P. Nanayakkara, H. Alper, Adv. Synth. Catal. 348 (2006) 545–550.

[2] J. Dheur, M. Sauthier, Y. Castanet, A. Mortreux, Adv. Synth. Catal. 352 (2010) 557–561.

[3] a. C.Y. Lin, Y.C. Cheng, A.I. Tsai, C.P. Chuang, Org. Biomol. Chem. 4 (2006) 1097−1103.
    b. K.P. Chen, H.Q. Lee, Y.C. Cheng, C.P. Chuang, Org. Biomol. Chem. 7 (2009) 4074−4081.

[4] Z.Y. Lin, Y.L. Chen, C.S. Lee, C.P. Chuang, Eur. J. Org. Chem. (2010) 3876−3882.

[5] F. Babudri, S.R. Cicco, G.M. Farinola, L.C. Lopez, F. Naso, V. Pinto, Chem. Commun. (2007) 3756−3758.

[6] A.N. Butkevich, L. Meerpoel, I. Stansfield, P. Angibaud, A. Corbu, J. Cossy, Org. Lett. 15 (2013) 3840−3843.

[7] Y. Wang, L. Xu, D. Ma, Chem. Asian J. 5 (2010) 74−76.

[8] L.B. Zhao, Z.H. Guan, Y. Han, Y.X. Xie, S. He, Y.M. Liang, J. Org. Chem. 72 (2007) 10276−10278.

[9] S. Kramer, J.L.H. Madsen, M. Rottländer, T. Skrydstrup, Org. Lett. 12 (2010) 2758−2761.

CHAPTER 5

Synthesized by Intramolecular Cyclizations

Among all the procedures for furan synthesis, intramolecular cyclization represents the most straightforward pathway.

In the numerous known methodologies for furan synthesis via intramolecular cyclization, enynols were studied with various transition metal catalysts. As early as in 1994, Bruneau, Dixneuf, and their co-workers explored the application of ruthenium catalysts on this topic [1]. The desired furans were obtained in moderate yields under neutral conditions by selective cyclization of (Z)-pent-2-en-4-yn-ols with $RuCl_2(PPh_3)(p$-cymene) (1 mol%) as the catalyst precursor. Later on, they prepared a class of new ruthenium complexes which are more electrophilic. These new complexes were applied in the cyclization of enynol as well, and it was found that improved yield can be achieved (Scheme 5.1).

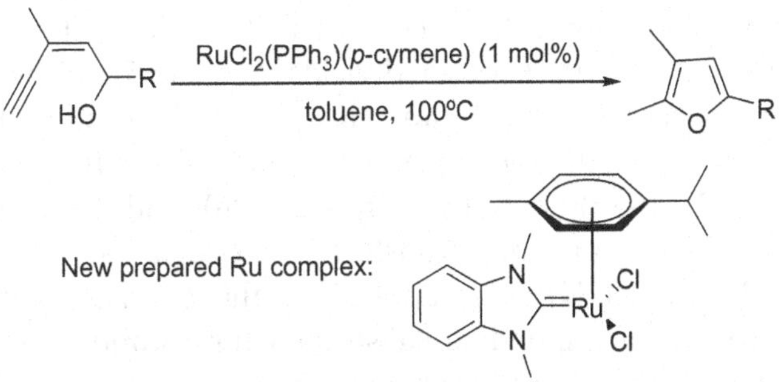

Scheme 5.1 Ru-catalyzed synthesis of furans from (Z)-pent-2-en-4-yn-ols.

In 2006, a report on applying trihydrazinophosphaadamantane THPA as a suitable water-soluble ligand for transition metal-catalyzed furan synthesis in aqueous phase was published [2]. Ruthenium, rhodium, and iridium complexes containing the trihydrazinophosphaadamantane ligand were prepared and studied for the first time. All the complexes prepared, which were soluble in water as well as in common organic solvents, showed a good activity in the catalytic isomerization

Transition Metal Catalyzed Furan Synthesis. DOI: http://dx.doi.org/10.1016/B978-0-12-804034-8.00005-2

of allylic alcohols into the corresponding ketones in both THF and water. These catalysts were found to be efficient for the cycloisomerization of (Z)-2-en-4-yn-1-ols to the corresponding functionalized furans in aqueous media as well. The best performances in terms of activity and recyclability (up to 10 cycles) have been obtained using the iridium catalyst which is the first example of an iridium-catalyzed cycloisomerization of (Z)-2-en-4-yn-1-ols.

$CuCl_2 \cdot 2H_2O$ was applied as catalyst on this transformation in 1990 [3]. In the presence of catalytic amount of copper salt, excellent yield of 2,3-dimethylfuran was prepared from (Z)-3-methyl-2-penten-4-yn-1-ol. In 2000, square planar cationic rhodium(I) dicarbonyl complexes, $[\{Rh((mim)_2CH_2)(CO)_2\}^+ BPh_4^-]$ and $[\{Rh((mBnzim)_2CH_2)(CO)_2\}^+ BPh_4^-]$ [mim = N-methylimidazol-2-yl, mBnzim = N-methylbenzimidazol-2-yl], were prepared and applied in cyclization reactions [4]. Lactones can be prepared from alkynoic acids, such as 4-pentynoic acid, 4-hexynoic acid, and 5-hexynoic acid. Cyclizations of acetylenic alcohols were carried out as well and it gave the corresponding oxygen-containing heterocycles in excellent yields.

In 1999, a general and facile palladium-catalyzed synthesis of furans from enynols was reported [5]. In the presence of K_2PdI_4 under essentially neutral conditions, (Z)-2-en-4-yn-1-ols were cyclized into the desired furans at $25-100°C$ in good yields (Scheme 5.2). The precursor for naturally occurring rosefuran was effectively produced as well. Later on, a new palladium complex was prepared, cis-[PdCl$_2$(κ^2-(P,N)-Ph$_2$PCH$_2$P$\{=$NP($=$O)(OR)$_2\}$Ph$_2$)] (R = Et, Ph), and it was found that improved activity for this transformation can be achieved [6]. Recently, $PdCl_2(MeCN)_2$ was applied in the cyclization of various Z-enynols to the corresponding substituted furans and dihydrofurans. The Z-enynols were produced via the reaction of vinyllithium intermediate with aldehydes and ketones [7].

Scheme 5.2

Reactions were carried out on a $3-10$ mmol scale based on (Z)-enynol. PdI_2 and KI (2 mol per mol of palladium) were added to pure (Z)-enynol or to a solution of (Z)-enynol in dry DMAc in a Schlenk flask. The resulting mixture was stirred at the room temperature and for the time required to obtain a satisfactory conversion, as shown by gas-liquid chromatography (GLC) and/or Thin Layer Chromatography (TLC) analysis.

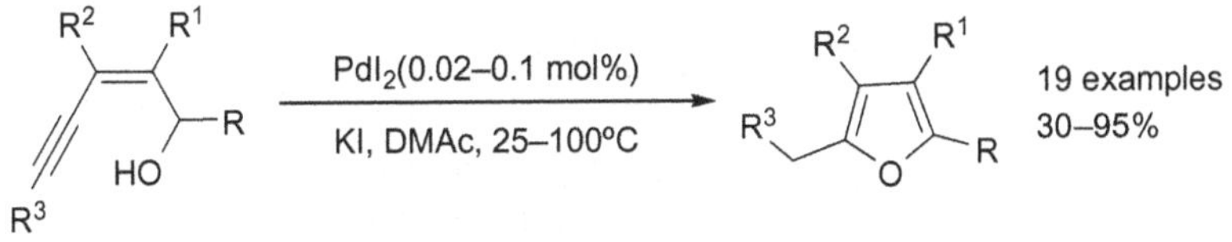

Scheme 5.2 Pd-catalyzed synthesis of furans from Z-enynols.

Liu and co-workers developed a gold-catalyzed cyclization of (Z)-enynols to the corresponding furans in 2005 [8]. The reaction proceeds under neutral conditions at room temperature and fully substituted furans were formed in high yields (Scheme 5.3). Both $AuCl_3$ and $(PPh_3)AuCl$ were found to be active for this transformation. Later on, $AuBr_3$ was studied and reported as well [9]. McDonald and co-workers found that molybdenum pentacarbonyl ($Mo(CO)_5$) together with NEt_3 can act as an effective promoter for the cyclization of enynols as well [10].

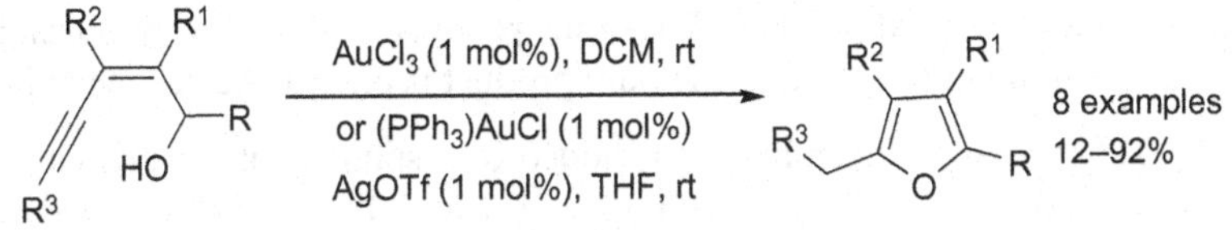

Scheme 5.3 Au-catalyzed synthesis of furans from Z-enynols.

In 1995, Marshall and Sehon reported a silver-catalyzed furan synthesis [11]. With alkynyl allylic alcohols, allenones, and allenylcarbinols as the substrates, the corresponding furans and 2,5-dihydrofurans can be produced in excellent yields (Scheme 5.4). Notably, in the optimization with β-alkynyl allylic alcohol, excellent yield can be achieved with all the tested silver salts and solvents at room temperature. This system was applied by Qing and co-workers in the cyclization of 2-alkynyl-3-trifluoromethyl allylic alcohols to produce the corresponding 3-trifluoroethylfurans [12]. In 2007, Yuan and Zhang reported the synthesis of α-(trifluoromethyl)furans from the corresponding (E)-3-aryl-2-(2-phenylethynyl)-1-(trifluoromethyl)allyl alcohols with AgOTf as the catalyst [13].

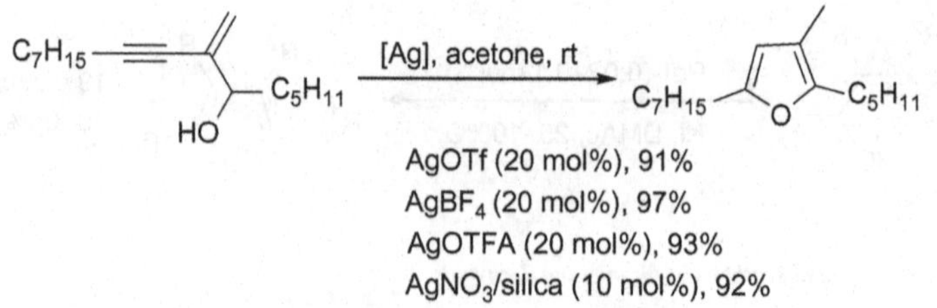

Scheme 5.4 Ag-catalyzed synthesis of furans from β-alkynyl allylic alcohol.

Lee and Kim reported the cyclization of allenyne-1,6-diols in 2008 [14]. With gold or silver as the catalysts, 2,5-dihydrofuran or furan derivatives can be selectively produced in good to excellent yields (Scheme 5.5). The difference in selectivity is due to the gold- or silver-catalyzed selective activation and differentiation of the double and triple bonds in allenyne-1,6-diols.

Scheme 5.5

AuCl$_3$ (9.8 mg, 0.032 mmol) was added to a solution of 1,6-diphenyl-2-vinylidenehex-3-yne-1,6-diol (63.0 mg, 0.217 mmol) in dichloromethane (1.5 ml). After the reaction mixture had been stirred for 10 min, the solvent was removed under reduced pressure. The crude product was purified by silica gel column chromatography to give the pure product using EtOAc/hexane = 1/5.

AgOTf (8.4 mg, 0.032 mmol) was added to a solution of 1,6-diphenyl-2-vinylidenehex-3-yne-1,6-diol (63.0 mg, 0.217 mmol) in dichloromethane (1.5 ml). After the reaction mixture had been stirred for 1.5 h, the solvent was removed under reduced pressure. The crude product was purified by silica gel column chromatography (EtOAc/hexane = 1/5) to give the pure product.

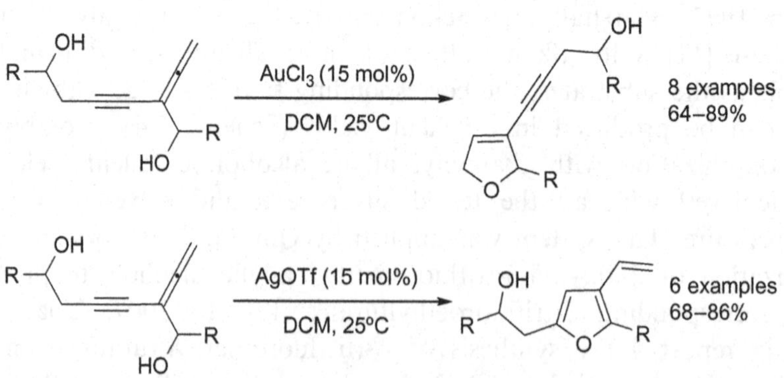

Scheme 5.5 Ag- and Au-catalyzed transformation of allenyne-1,6-diols.

Hashmi's group studied the gold(I)-catalyzed cyclization of 2-alkynylallyl alcohols in 2011 [15]. With their gold complexes, 2,4-disubstituted and 2,3,5-trisubstituted furans were prepared in good yields under mild conditions with broad scope (Scheme 5.6). Bisfurans can be produced as well from the corresponding bifunctional substrates.

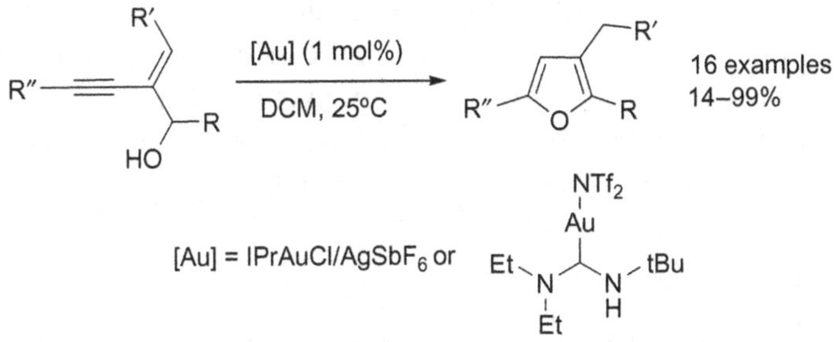

Scheme 5.6 Au-catalyzed transformation of 2-alkynylallyl alcohols.

Lu and co-workers developed a palladium-catalyzed procedure for the synthesis of 3,3,3-trifluoroprop-1-en-αyl-substituted furans [16]. With 1,1,1-trifluoro-2-[(*tert*-butyldimethylsilyloxy)methyl]-3-alkynylbut-2-en-1-ols, which were obtained from 1,1,1-trifluoro-2-[(*tert*-butyldimethyl-silyloxy)methyl]-3,3-dibromoprop-2-ene in three steps, as the substrates, in the presence of palladium catalyst, excellent yields of the desired furans can be isolated (Scheme 5.7).

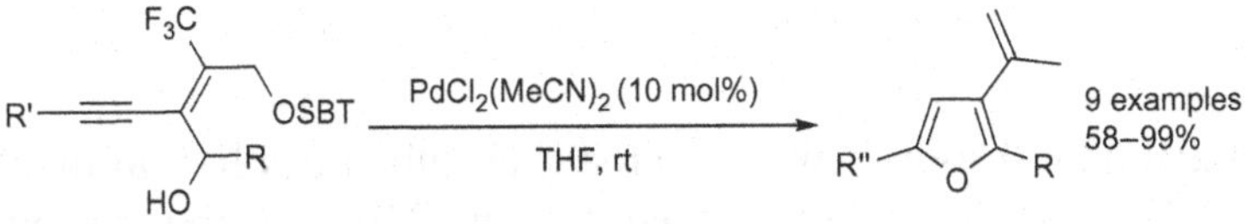

Scheme 5.7 Pd-catalyzed synthesis trifluoro-substituted furans.

Gabriele et al. reported a palladium-catalyzed carbonylative synthesis of furan-2-acetic esters from (Z)-2-en-4-yn-1-ols in 1997 [17]. In the presence of catalytic amount of PdI_2 and KI, the desired furans were obtained in good yields at $50-70°C$ and under 100 bar of a 9:1 mixture of carbon monoxide and air (Scheme 5.8a). Later on, they developed a procedure for the synthesis of 2-furan-2-ylacetamides by sequential palladium-catalyzed oxidative aminocarbonylation of (Z)-2-en-4-yn-1-ols/conjugate addition/aromatization (Scheme 5.8b) [18]. The method is based on a PdI$_2$-catalyzed oxidative aminocarbonylation of the triple bond of (Z)-2-en-4-yn-1-ols to give the corresponding 2-ynamide intermediates, which undergo intramolecular conjugate addition to give 2-(5H-furan-2-ylidene)acetamide derivatives. Spontaneous or one-pot acid-promoted aromatization of 2-(5H-furan-2-ylidene)acetamides eventually leads to the final furanacetamide derivatives. Recently, they developed the carbonylative synthesis of furan-3-carboxylic esters from 3-yne-1,2-diols (Scheme 5.8c) [19]. The process consisted of a sequential combination between a 5-endo-dig heterocyclo dehydration step and a mild palladium-catalyzed oxidative alkoxycarbonylation stage (100°C in ROH under 40 bar of a 4:1 mixture of CO-air).

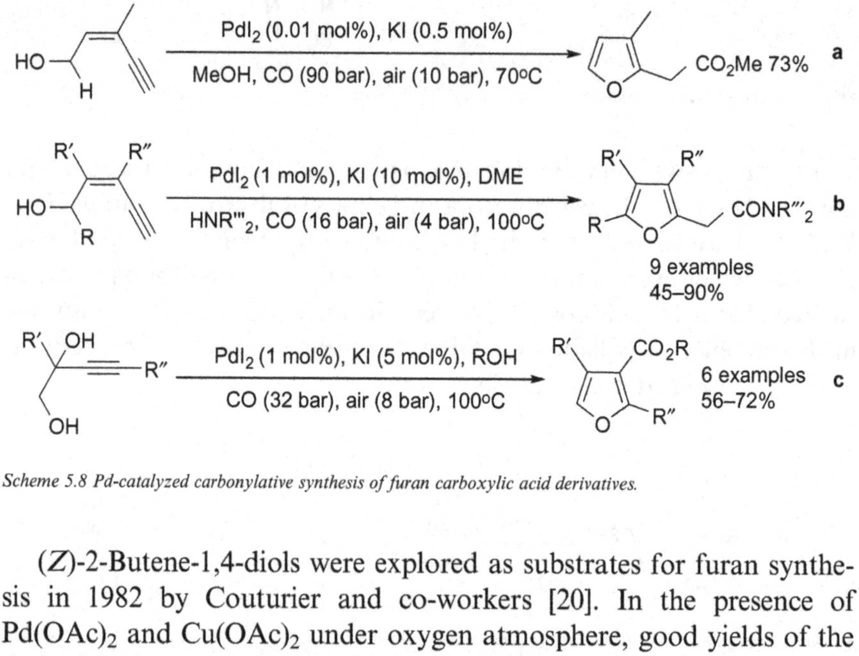

Scheme 5.8 Pd-catalyzed carbonylative synthesis of furan carboxylic acid derivatives.

(Z)-2-Butene-1,4-diols were explored as substrates for furan synthesis in 1982 by Couturier and co-workers [20]. In the presence of Pd(OAc)$_2$ and Cu(OAc)$_2$ under oxygen atmosphere, good yields of the desired furans can be isolated (Scheme 5.9).

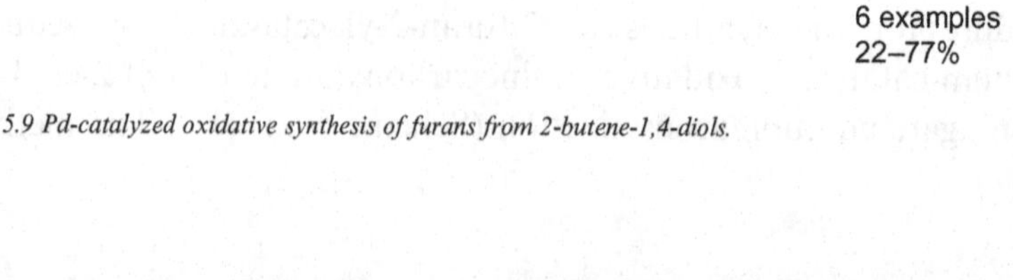

Scheme 5.9 Pd-catalyzed oxidative synthesis of furans from 2-butene-1,4-diols.

Lu and Ji developed a procedure for the synthesis of 2,5-disubstituted furans from alkynediols in 1993 [21]. By this isomerization and dehydration sequence, with $Pd_2(dba)_3 \cdot CHCl_3$, nBu_3P, and perfluorinated resin-sulfonic acid as the catalyst system at 130°C, 2,5-disubstituted furans were formed in good yields (Scheme 5.10).

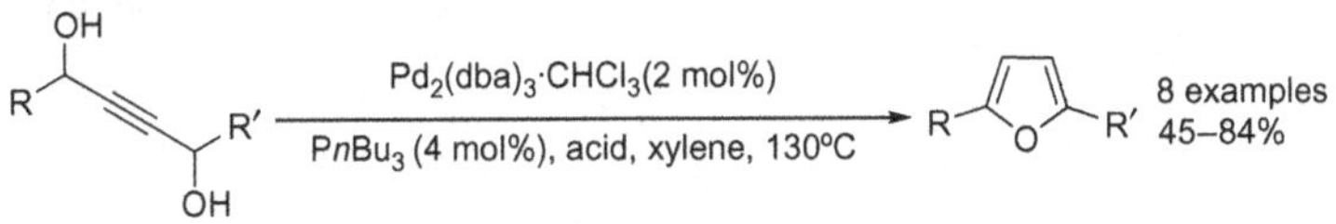

Scheme 5.10 Pd-catalyzed oxidative synthesis of furans from alkynediols.

Williams and co-workers established a ruthenium-catalyzed procedure for the synthesis of furans from alkynediols [22]. Using ruthenium dihydride $[Ru(PPh_3)_3(CO)H_2]$ and Xantphos as the catalyst system with an organic acid co-catalyst, various furans could be made in one tandem reaction in good yields (Scheme 5.11). Concerning the reaction pathway, the alkynediols were transformed into the corresponding 1,4-diketones which then undergo a Paal-Knorr cyclization to give the final furans. This transformation was also described with $[Rh(rac\text{-}BINAP)]BF_4$ catalyst by Tanaka's group as well in their study on the isomerization of secondary propargylic alcohols to α,β-enones [23].

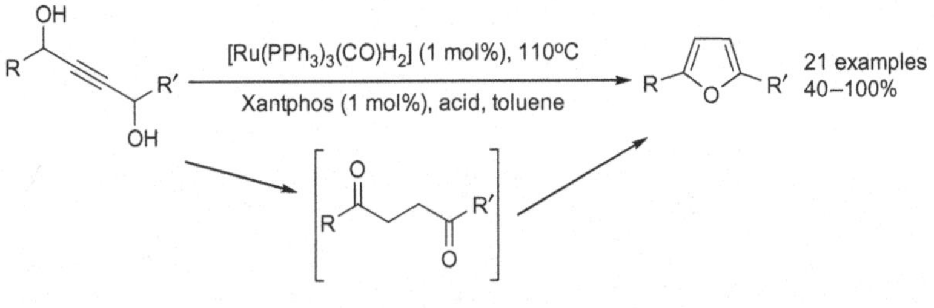

Scheme 5.11 Ru-catalyzed oxidative synthesis of furans from alkynediols.

Knight and co-workers reported a heterogeneous system for the synthesis of furans from diols [24]. A wide variety of 3-alkyne-1,2-diols were undergone 5-endo-dig cyclizations followed by dehydration at ambient temperature and gave the corresponding furans in essentially quantitative yields with 10 mol% of 10% w/w silver(I) nitrate absorbed on silica gel as the catalyst (Scheme 5.12). This methodology was also applied in the total synthesis of plakorsin B by the same group [25].

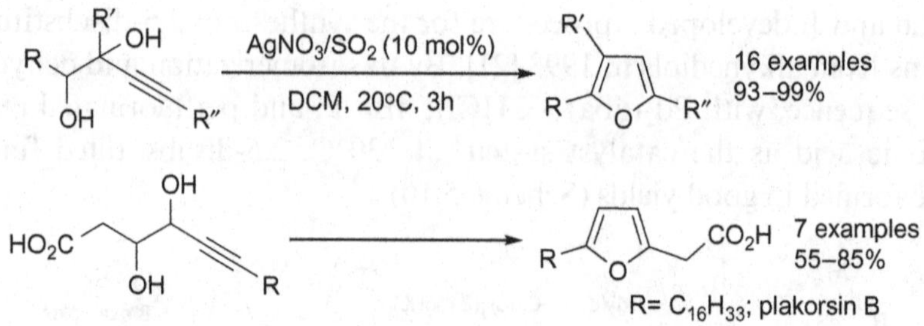

Scheme 5.12 Ag/SO$_2$-catalyzed oxidative synthesis of furans from 3-alkyne-1,2-diols.

Lee and co-workers developed a gold-catalyzed procedure for transforming enyne-1,6-diols into the corresponding furans in 2010 [26]. In the treatment of enyne-1,6-diols with 5 mol% Ph$_3$PAuCl in the presence of 5 mol% AgOTf as the co-catalyst, trisubstituted furans were selectively produced in good to excellent yields in dichloromethane at room temperature for 5–10 min through cyclization followed by isomerization (Scheme 5.13).

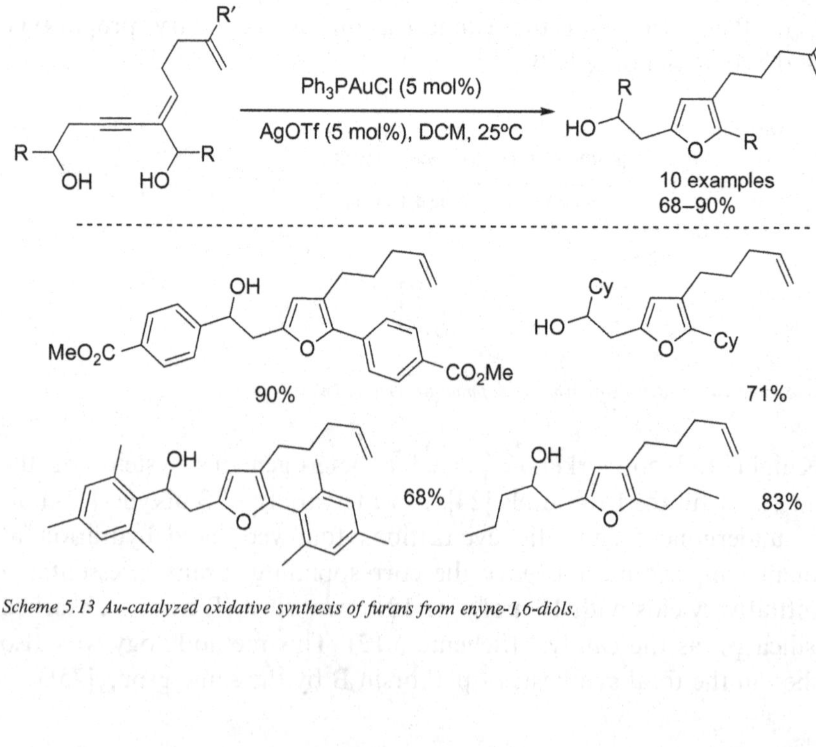

Scheme 5.13 Au-catalyzed oxidative synthesis of furans from enyne-1,6-diols.

A simple copper-catalyzed procedure for the synthesis of substituted furans and pyrroles was developed by Gabriele and co-workers in 2010 [27]. With CuCl$_2$ as the catalyst via heterocyclodehydration of readily available 3-yne-1,2-diols and N-Boc- or N-tosyl-1-amino-3-yn-2-ols, in MeOH at 80–100°C for 1–24 h, the corresponding heterocyclic derivatives were formed in 53–99% isolated yields (Scheme 5.14).

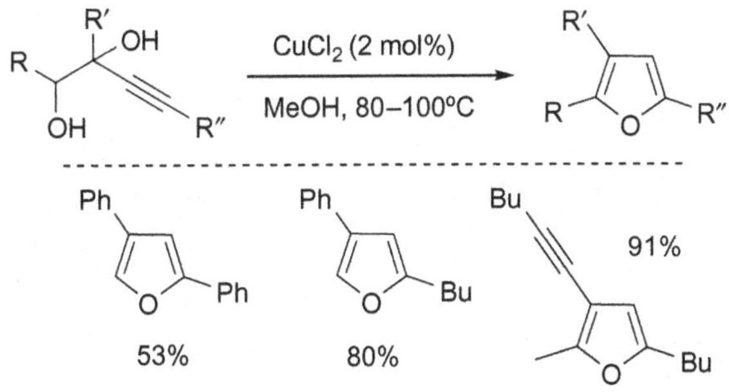

Scheme 5.14 Cu-catalyzed oxidative synthesis of furans from 3-alkyne-1,2-diols.

Ma and co-workers developed a procedure for the synthesis of polysubstituted furans in 2010 [28]. The reaction is based on a stepwise Sonogashira coupling of (Z)-3-iodoalk-2-en-1-ols with terminal propargylic alcohols and subsequent Au(I)- or Pd(II)-catalyzed cyclization-aromatization via elimination of H$_2$O. The Sonogashira coupling of 3-iodoalk-2-en-1-ols with terminal propargyl alcohols gives 4-alkyn-2-ene-1,6-diols as the intermediates. Subsequent cycloisomerization in DMAc or CH$_2$Cl$_2$ with Au(PPh$_3$)Cl and AgOTf as the catalyst would afford polysubstituted 2-(1-alkenyl)furans; with PdCl$_2$ as the catalyst and the reaction in DMAc in the presence of allylic bromides, the same substrates afforded polysubstituted 2-(1,4-alkadienyl) furans. In both types of catalyzed cyclization reactions, the elimination of H$_2$O promoted the aromatization to form the furan ring. Different alkyl or aryl groups could be introduced into different positions of furans due to the substituent-loading capability of 3-iodoalkenols and diversity of the terminal propargyl alcohols and allylic bromides (Scheme 5.15).

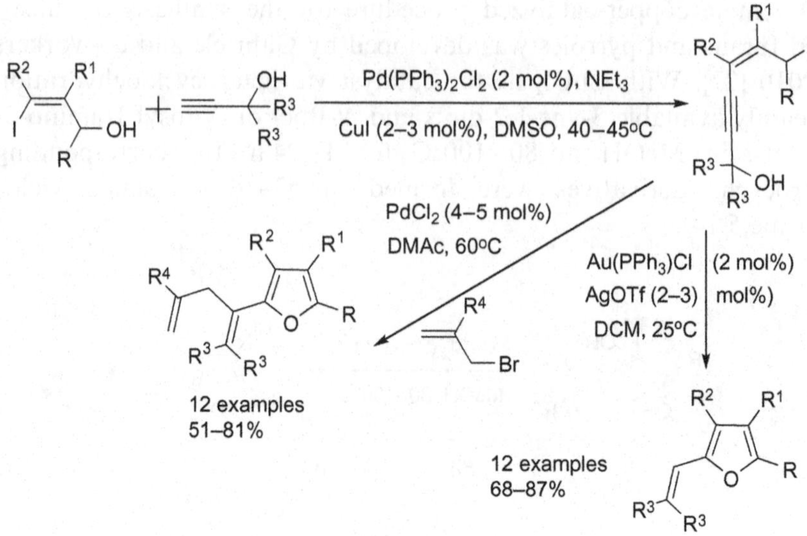

Scheme 5.15 Cascade synthesis of furans from (Z)-3-iodoalk-2-en-1-ols.

Palladium-catalyzed heterocyclization-coupling sequences with buta-1,2,3-trienyl carbinols and electron-deficient alkenes as substrates to the corresponding furans were reported as well in 2008 [29]. Polysubstituted furans were formed where the heterocyclic ring originates from the elements of the butatrienyl carbinol while the electron deficient olefin is incorporated as a C-3 substituent (Scheme 5.16). In most cases, the reaction proceeds via a Heck-type pathway leading to the efficient formation of 3-vinylfurans. However, couplings with methyl vinyl ketone display a divergent behavior to afford selectively either Heck- or hydroarylation-type products depending on reaction conditions.

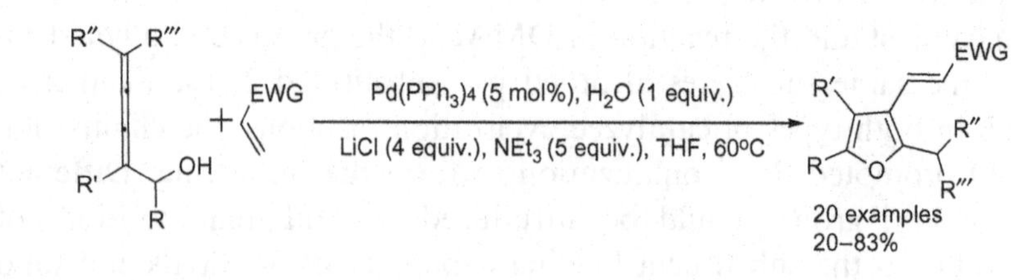

Scheme 5.16 Cascade synthesis of furans from buta-1,2,3-trienyl carbinols.

A palladium-mediated three-component coupling reaction between 2-iodotoluene, 1-penten-4-yn-3-ol, and diethyl ethoxymethylenemalonate was reported by Morimoto and co-workers in 2005 [30]. This procedure was applied in the total synthesis of 13-hydroxy-14-nordehydrocacalohastine and 13-acetoxy-14-nordehydrocacalohastine, two novel modified furanoeremophilane-type sesquiterpenes isolated from *Trichilia cuneata*, which showed inhibitory activities for membrane lipid peroxidation in mitochondria and microsomes (Scheme 5.17).

Scheme 5.17

n-Butyllithium (12.5 ml, 20.0 mmol, 1.6 M in hexane) was added dropwise to a solution of alcohol (1.64 g, 20.0 mmol) in tetrahydrofuran (12.5 ml) at 0°C under a nitrogen atmosphere, and the solution was allowed to reach room temperature (15 min). In a separate flask, *n*-butyllithium (0.625 ml, 1.0 mmol) was added dropwise to a well-stirred suspension of dichlorobis(triphenylphosphine)palladium(II) (350 mg, 0.5 mmol) in dimethyl sulfoxide (20 ml) (rt, 1 h) to give a dark red homogeneous solution. To this palladium solution were successively added dropwise a mixture of diethyl ethoxymethylenemalonate (2.02 ml, 10.0 mmol) and 2-iodotoluene (1.08 ml, 10.0 mmol) in tetrahydrofuran (10 ml), and the lithium alkoxide solution at room temperature. The reaction mixture was stirred at 40°C for 12 h. A saturated aqueous solution of ammonium chloride (20 ml) and water (50 ml) was added to the reaction mixture, and the resulting mixture was filtered through a pad of Celite under reduced pressure. The filtrates were extracted with ether (40 ml × 3). The organic layer was washed with brine, dried over anhydrous sodium sulfate, and concentrated in vacuo. The residue was purified by column chromatography (hexane/ethyl acetate = 50:1) on 270 g of silica gel to yield products.

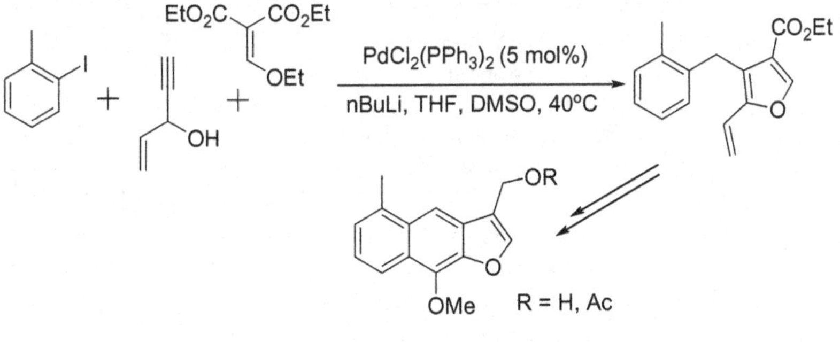

Scheme 5.17 Cascade synthesis of furans from 1-penten-4-yn-3-ol.

Nishibayashi and co-workers developed a ruthenium-catalyzed intramolecular cyclization of 3-butyne-1,2-diols into the corresponding substituted furans in 2008 [31]. In the presence of $[Cp^*RuCl(\mu^2\text{-}SMe)]^2$

as the catalyst, good to high yields of the corresponding furans can be achieved (Scheme 5.18). This catalytic reaction was proposed to proceed via ruthenium-allenylidene complexes as the key intermediates. When using 1-amino-3-butyn-2-ol as the substarte at 70°C for 24 h, 2-phenylpyrrole can be formed in 55% yield.

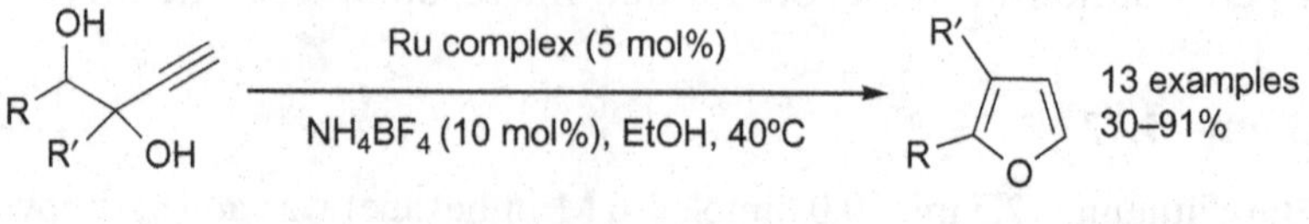

Scheme 5.18 Ru-catalyzed synthesis of furans from 3-butyne-1,2-diols.

In 2009, Akai's research group reported a gold-catalyzed version of this transformation [32]. With a low catalyst loading (0.05–0.5 mol%) of $(Ph_3P)AuCl$-$AgNTf_2$ or $(Ph_3P)AuCl$-$AgOTf$ at room temperature, a variety of substituted furans and pyrroles were formed in excellent yields (85–98% yields) from the corresponding 3-alkyne-1,2-diols and the 1-amino-3-alkyn-2-ols via intramolecular cyclizations (Scheme 5.19a). This method is also fully applicable to the conversion of several dozen grams of the substrate using only 0.05 mol% each of the Au and Ag catalysts. Later on, this methodology was applied by Hong and co-workers in the total synthesis of (±)-cafestol [33]. Meanwhile, Aponick and co-workers reported the usage of AuCl as catalyst for the same transformation (Scheme 5.19b) [34]. In this procedure, thiophenes can be produced as well.

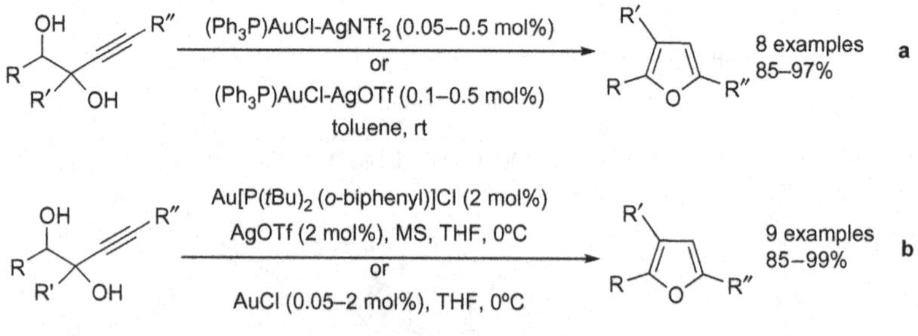

Scheme 5.19 Au-catalyzed synthesis of furans from 3-butyne-1,2-diols.

Homopropargylic alcohols were applied as substrates in furan synthesis under the assistant of platinum catalyst in 2011 [35]. In this procedure, homopropargylic alcohols bearing a suitable leaving group can be rearranged to form α,β-unsaturated carbenes. These intermediate species were subsequently converted to polysubstituted furans in short reaction times with excellent overall yields (Scheme 5.20).

Olefin-based ligands, ethylene in particular, were found to be the best ligand for this reaction. The cycloisomerization conditions are significantly mild, without air or moisture sensitivity, and the reaction shows a wide array of functional group compatibility.

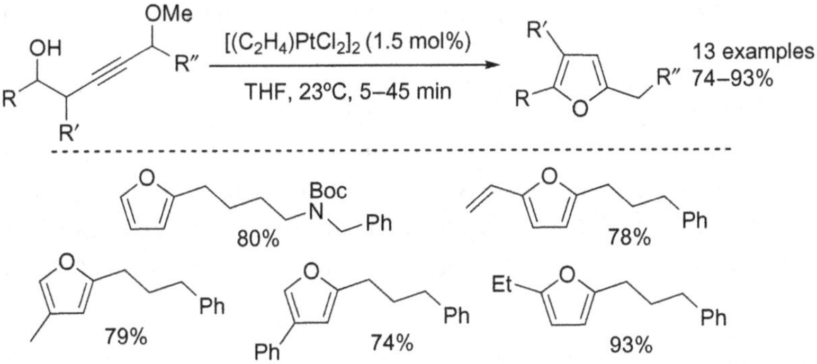

Scheme 5.20 Pt-catalyzed synthesis of furans from homopropargylic alcohols.

Recently, a mercury-catalyzed synthesis of furans from alkyne diols was reported [36]. Spiroketals were constructed from alkyne diols effectively in aqueous conditions with Hg(II) salts as the catalyst. Monounsaturated spiroketals and furans were accessed with equal ease when propargylic triols (or propargylic diols) were subjected to similar conditions (Scheme 5.21). The reactions were instant and high-yielding at ambient temperatures. Regioselectivity issues are well addressed as well. Mercury catalyst was also applied in the cyclization of allenic substrates which *in situ* produced from 3-methoxy-1-phenylthio-1-propyne and aldehydes [37]. Good yields of 3-phenylthio and 3-methoxy substituted furans can be produced.

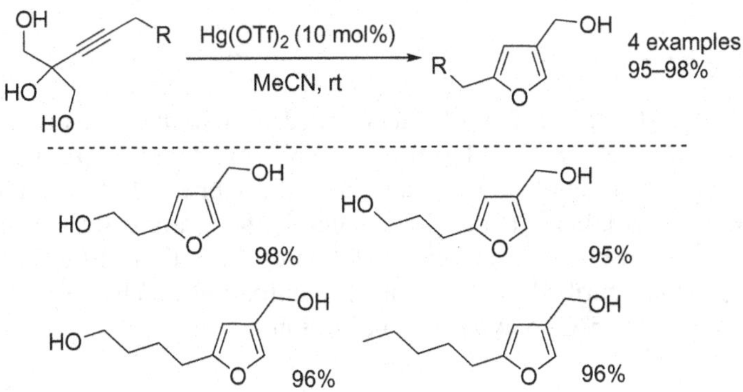

Scheme 5.21 Hg-catalyzed synthesis of furans from alkyne diols.

Palladium catalyst was applied in the cyclization of 3-alkyn-1,2-diols and 2-methoxy-3-alkyn-1-ols to the corresponding furans as early as in 1985 [38]. The intermediary 3-furylpalladiums can be trapped with allyl halides and gave the corresponding 3-allylfurans in good yields. Silver salt ($AgNO_3$) was applied as catalyst for this type of transformation as well, and applied in the synthesis of zoanthamine alkaloids [39].

An interesting method for the synthesis of highly substituted iodine-containing furans was developed by Liang and co-workers [40]. The iodine-containing furans were produced by cyclization of 1-(arylethynyl)-2,3-epoxy alcohols with alcohols as nucleophiles in the presence of iodide under very mild reaction conditions in good yields (Scheme 5.22).

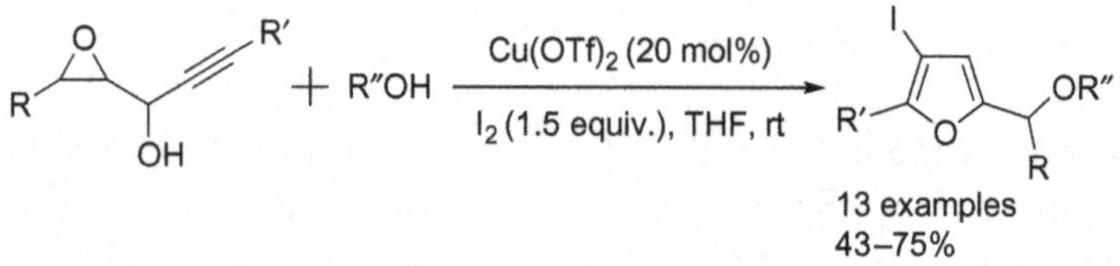

Scheme 5.22 Cu-catalyzed synthesis of furans from 1-(arylethynyl)-2,3-epoxy alcohols.

A gold-catalyzed procedure for the synthesis of 3-alkoxyfurans was reported in 2014 by Sheppard and co-workers [41]. By treatment of acetal-containing propargylic alcohols with 2 mol% gold catalyst in an alcohol solvent at room temperature, the desired furans were formed in excellent yields (Scheme 5.23).

Scheme 5.23

[Ph$_3$PAuNTf$_2$]$_2$·PhMe (2 mol%) was added to a solution of propargylic alcohol in alcohol (8–10 ml/g) and the solution stirred magnetically at room temperature until starting material had disappeared (TLC). The solvent was removed in vacuo (cold water bath—heating during solvent evaporation can promote aerobic oxidation of the 3-alkoxyfuran) and the crude product purified by column chromatography. The product was then stored at 0–5°C to avoid decomposition.

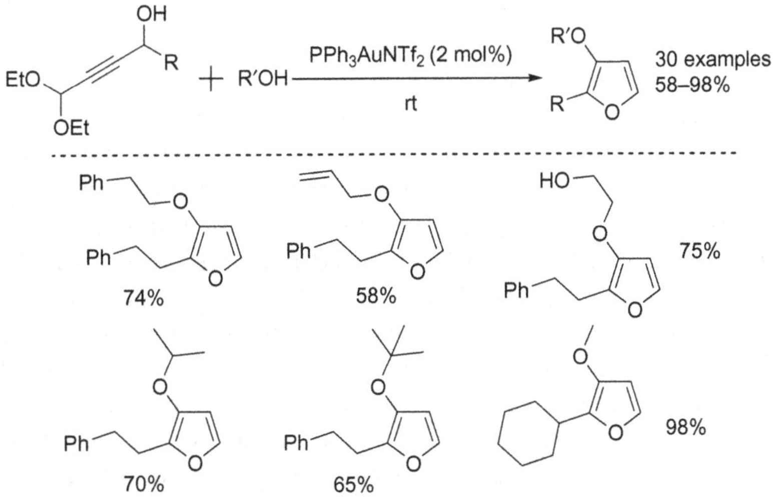

Scheme 5.23 Au-catalyzed synthesis of 3-alkoxyfurans.

Acetylenic α,β-epoxides can be applied as substrates for furan synthesis as well [42]. Under the assistant of dilute sulphuric acid and mercuric sulphate on heating, furans can be produced in good yields. The reaction involves an internal hydration at a terminal acetylenic carbon atom. Under the influence of sulphuric acid alone, only small quantities of furans were formed; the major products were acetylenic diols. When the epoxides treated with dilute sulphuric acid in the presence of ethanol, monoethers of acetylenic diols were formed. Acetylenic β,γ-epoxides have also been shown to give furans in the mercuric sulphate-catalyzed reaction with sulphuric acid.

Later on, a mercury-catalyzed procedure based on 1-alkynyl-2,3-epoxyalcohols was developed by Marson and co-workers [43]. 2,3,5-Trisubstituted furans can be formed in good yields by the reaction of 1-alkynyl-2,3-epoxyalcohols with a catalytic amount of mercury(II) prepared from HgO and dilute sulfuric acid (Scheme 5.24).

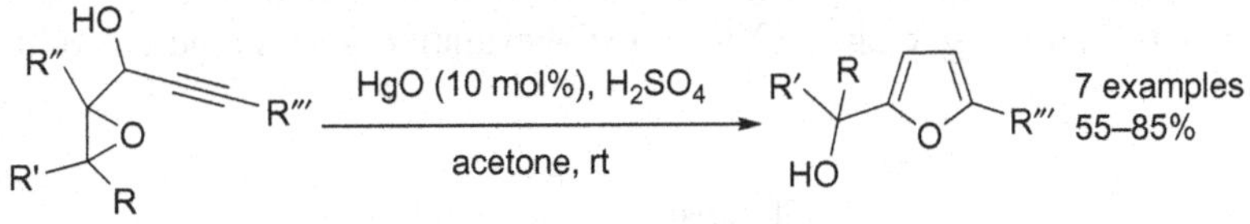

Scheme 5.24 Hg-catalyzed synthesis of furans from 1-alkynyl-2,3-epoxyalcohols.

In 2001, Aurrecoechea's group reported the synthesis of trisubstituted furans with epoxypropargyl esters as substrates [44]. By one-pot
sequential SmI$_2$-promoted reduction-elimination and Pd(II)-catalyzed
cycloisomerization, good yields of the furans can be achieved. The
sequence was initiated by a SmI$_2$-promoted reduction that takes advan-
tage of the ability of the alkynyloxirane moiety present in substrates to
accept electrons from SmI$_2$. The resulting organosamarium species then
eliminated an adjacent acetate- or benzoate-leaving group, leading to
the formation of unstable 2,3,4-trien-1-ols. Without isolation, these
were cycloisomerized to furans by treatment with a catalytic amount of
a Pd(II) complex and a proton source. The whole sequence takes place
under mild reaction conditions. Some useful functional groups such as
cyano and α,β-unsaturated esters can be tolerated, but benzyl- and
silyl-protected hydroxyl groups were deprotected to some extent.
Starting materials can be easily assembled using reliable reactions from
acetylene, an aldehyde or ketone, and a vinyl halide fragment.

Hashmi and co-workers reported a gold(III) chloride-catalyzed
furan synthesis in 2004 [45]. By the isomerization of alkynyl epoxides,
the desired furans were formed in good yields under mild conditions
(Scheme 5.25). Additional functional groups like hydroxyl groups or
aryl bromides can be tolerated in this procedure. For the mechanism,
it starts with the coordination of gold to the carboncarbon triple bond
and activates it for the addition of a nucleophile. Then the epoxideoxy-
gen would attack at the distal position of the alkyne, and the gold will
form a δ-bond to the other carbon atom of the alkyne and resulting
aromatic furan after loosing a proton. The final product can be produced after proto-demetallation.

Scheme 5.25

A solution of AuCl$_3$ in MeCN (66.1 mg; 5% w/w; 3.30 mg, 10.8 μmol
AuCl$_3$) was added to a solution of substrate (214 μmol) in MeCN (0.5 ml)
at 25°C. After 17 h (the reaction was monitored by thin layer chromatography) the solvent was evaporated and the residue was purified by column
chromatography (silica gel, 12% EtOAC-hexane) to give the pure product.

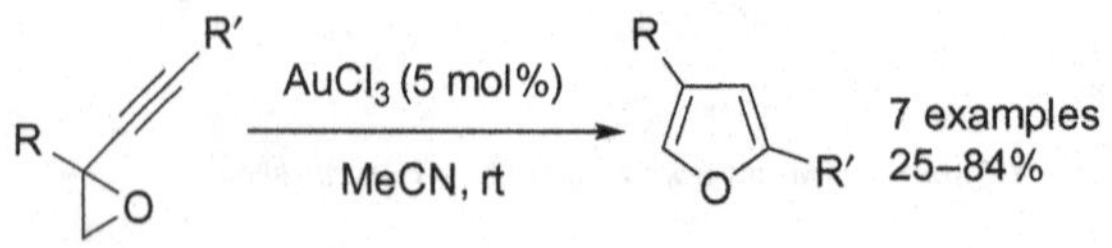

Scheme 5.25 Au-catalyzed synthesis of furans from alkynyl epoxides.

A gold-catalyzed tandem cycloisomerization of alkynyloxiranes with nucleophiles to the corresponding 2,5-disubstituted furans was reported by Liang and co-workers in 2007 [46]. The reaction proceeds efficiently under mild conditions with commercially available catalysts to afford furans in moderate to excellent yields (up to 96%) with high diversity (Scheme 5.26a). Later on, they found that difurylmethane derivatives can be produced by slightly changing the reaction conditions [47]. In the presence of 2 mol% of tetrachloroauric acid, tetrahydrate ($HAuCl_4 \cdot 4H_2O$), under very mild conditions, good yields of the products can be formed by cycloisomerization of 1-oxiranyl-2-alkynyl esters and dimerization of the 2-(α-hydroxyalkyl)furans (Scheme 5.26b). Additionally, they succeeded in combining their gold-catalyzed tandem cyclization with Friedel–Crafts-type reactions and gave new furan derivatives (Scheme 5.26c) [48].

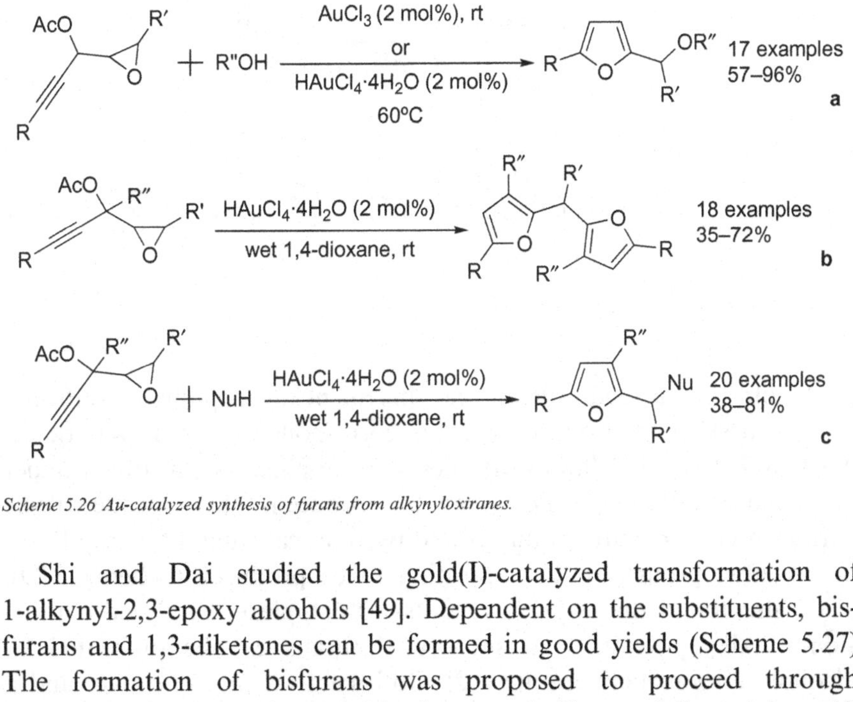

Scheme 5.26 Au-catalyzed synthesis of furans from alkynyloxiranes.

Shi and Dai studied the gold(I)-catalyzed transformation of 1-alkynyl-2,3-epoxy alcohols [49]. Dependent on the substituents, bisfurans and 1,3-diketones can be formed in good yields (Scheme 5.27) The formation of bisfurans was proposed to proceed through the sequential formation of 2-hydroxymethylfuran, followed by self-condensation in the presence of gold complex. Whereas the formation of 1,3-diketones was resulted from a domino C–C bond cleavage of epoxide system with the assistance of hydroxyl group and subsequent hydrolysis. Substituents on the oxirane have a significant effect on the selective formation of the two kinds of products.

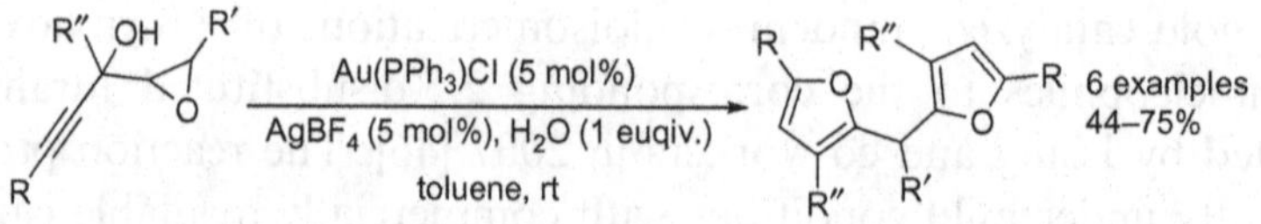

Scheme 5.27 Au-catalyzed synthesis of bisfurans from epoxides.

A regioselective gold-catalyzed cycloisomerization reaction of boron-containing alkynyl epoxides toward C2- and C3-borylated furans was developed by Gevorgyan and co-workers recently (Scheme 5.28) [50]. It was found that the copper catalyst as well as the gold catalyst with more basic triflate counterion favor boryl migration toward C3-borylated furans, whereas employment of the cationic gold hexafluoroantimonate affords C2-borylated furan via a formal 1,2-hydrogen shift.

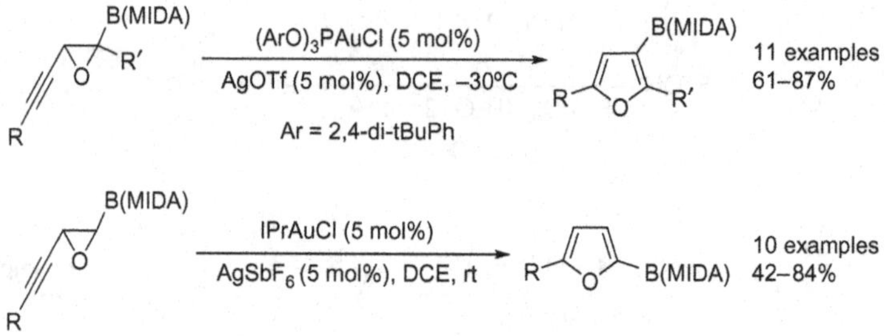

Scheme 5.28 Au-catalyzed synthesis of borylated furans from epoxides.

A gold(I)-catalyzed tandem rearrangement-nucleophilic substitution of α-acetoxy alkynyl oxiranes to the corresponding furans was developed in 2010 [51]. When aziridines were applied as substrates under the same conditions, pyrroles can be produced. This cascade allows various highly substituted furans and pyrroles carrying different alkoxy or alkylthio groups at the α-position to be produced (Scheme 5.29). They also found that Ag or Au salts- or complexes-catalyzed conversion of alkynyloxiranes to substituted furans can be promoted by alcohol at room temperature [52]. Both catalysts are effective, and a large number of furan diversity can be obtained in high yield with one or the other catalyst. Depending on the catalyst, one or both of the latter cyclized to dihydrofurans, and further elimination of the alcohol led to the corresponding furans. These results highlight the duality between oxophilicity and alkynophilicity of Ag or Au salts.

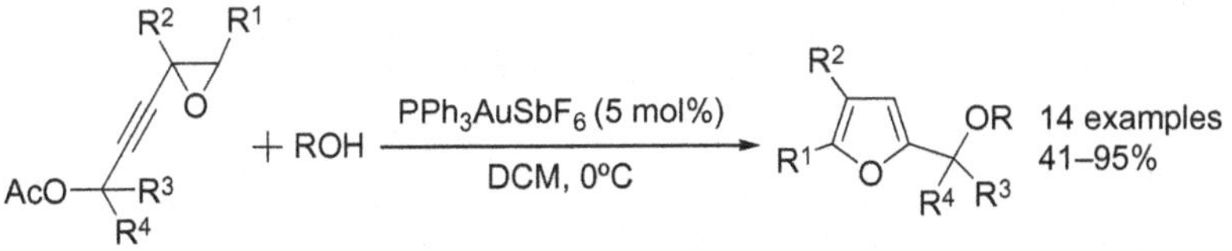

Scheme 5.29 Au-catalyzed synthesis of furans from oxiranes.

Pale and co-workers reported a silver(I)-catalyzed cascade to produce furans from alkynyloxiranes in 2009 [53]. Functionalized furans were conveniently formed by silver(I)-catalyzed reaction of alk-1-ynyl oxiranes in the presence of *p*-toluenesulfonic acid and methanol (Scheme 5.30). For the reaction mechanism, the starting alkynyloxirane probably was first opened by an alcohol in a nucleophilic substitution promoted by the catalytic amounts of *p*-toluenesulfonic acid. Then the formed β-alkoxy-β-alkynol would then cyclize upon coordination with silver ion, leading to heterocyclic vinyl silver species. Protodemetalation then regenerated the catalyst and gave 4-alkoxy-4,5-dihydrofuran. Upon elimination of alcohol, the latter would be converted to a furan derivative. This elimination could be facilitated either by the acid present or by the mild Lewis acid silver ion.

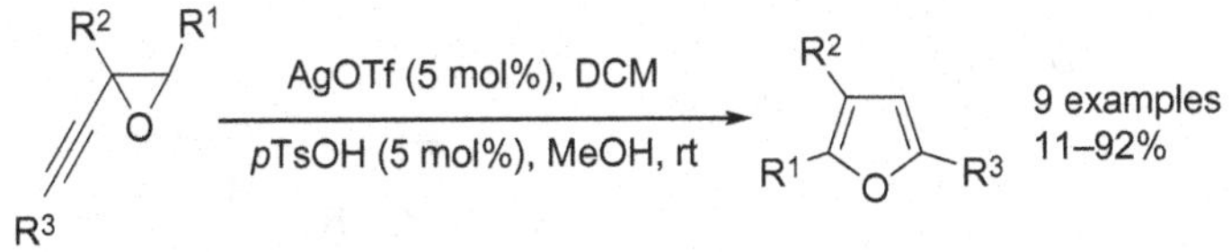

Scheme 5.30 Ag-catalyzed synthesis of furans from oxiranes.

Platinum catalyst was applied in the transformation of oxiranes to furans as well. In 2008, Yoshida and co-workers developed a methodology for the synthesis of substituted furans using a platinum catalyst from propargylic oxiranes in aqueous media [54]. A variety of substituted furans were produced in moderate to excellent yields (Scheme 5.31).

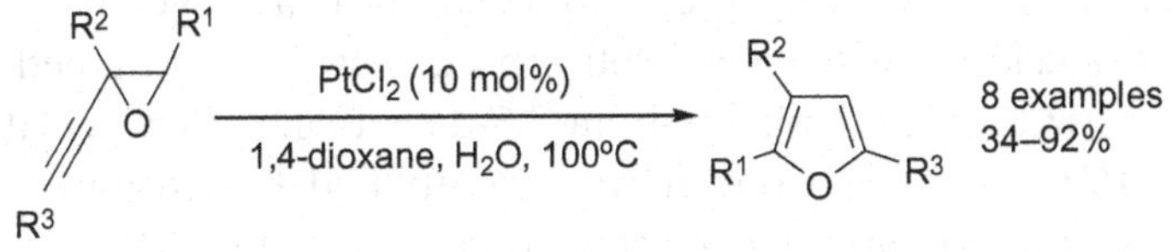

Scheme 5.31 Pt-catalyzed synthesis of furans from oxiranes.

Liu's research group reported a ruthenium-catalyzed cyclization of epoxyalkynes to furans in 2002 [55]. $TpRuPPh_3(CH_3CN)_2Cl$ together with NEt_3 was found to be the best system. The reactions worked well for various epoxyalkynes with suitable oxygen and nitrogen functionalities with low loading of catalyst (Scheme 5.32). It failed with disubstituted epoxyalkynes. A ruthenium-vinylidenium intermediate should be involved during the reaction and proven by deuterium-labeling experiment.

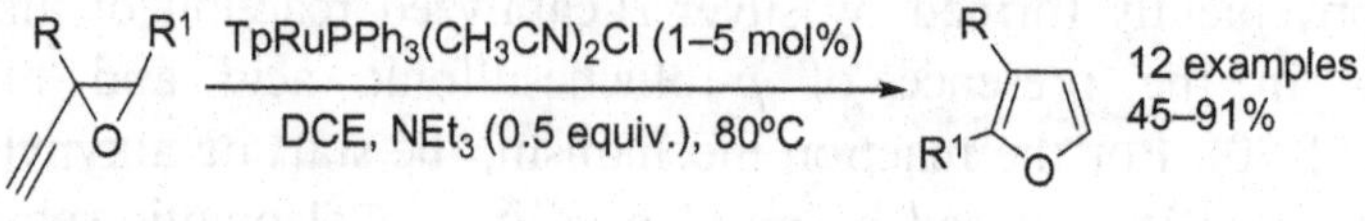

Scheme 5.32 Ru-catalyzed synthesis of furans from oxiranes.

Recently, a indium-catalyzed rearrangement of acetylenic epoxides to 2,3,5-trisubstituted furans was reported by Connell and Kang (Scheme 5.33) [56]. The epoxides applied here were directly prepared by nucleophilic ring closure of propargylic alkoxides generated by lithium acetylide addition to α-haloketones.

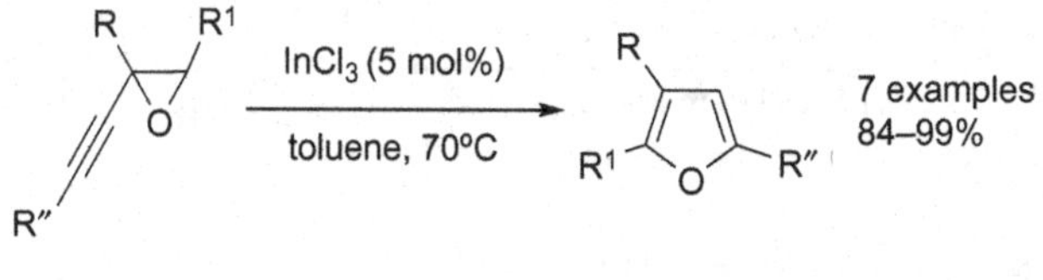

Scheme 5.33 In-catalyzed synthesis of furans from oxiranes.

Che and co-workers reported a gold(III) porphyrin-catalyzed cycloisomerization of allenones in 2005 [57]. With porphyrins as ligands, the Au(III) showed high reactivities and gave the corresponding furans in good to excellent yields (up to 98%) and with quantitative substrate conversions. By recovering the Au(III) catalyst, a recyclable catalytic system is developed with over 8300 product turnovers attained for the cycloisomerization of 1-phenyl-buta-2,3-dien-1-one. The versatility of the gold(III) porphyrin catalyst was exemplified by its application to the hydroamination and hydration of phenylacetylene in 73% and 87% yield, respectively.

Interestingly, Gevorgyan's group demonstrated a gold(III)-catalyzed 1,2-iodine, -bromine, and -chlorine migration in haloallenyl ketones for the synthesis of the corresponding 2-halofurans [58]. The reaction proceeded via a halirenium intermediate, and the desired products were formed in good yields (Scheme 5.34). In their mechanistic investigations through Density functional theory (DFT) computational and experimental studies, they found that both Au(I) and Au(III) catalysts activate the distal double bond of the allene to produce cyclic zwitterionic intermediates, which undergo a kinetically favored 1,2-Br migration [59]. However, in the cases of $Au(PR^3)L$ (L = Cl, OTf) catalysts, the counterion-assisted H-shift is the major process, indicating that the regioselectivity of the Au-catalyzed 1,2-H versus 1,2-Br migration is ligand-dependent. A computational study on Au-catalyzed cycloisomerizations proceeding via 1,2-Si or 1,2-H migrations was studied as well [60]. Valuable 2- and 3-silylfurans can be produced through a common Au-carbene intermediate. Both experimental and computational results clearly indicate that the 1,2-Si migration is kinetically favored over the 1,2-shifts of H, alkyl, and aryl groups in the -Si-substituted Au-carbenes. In addition, experimental results on the Au(I)-catalyzed cycloisomerization of homopropargylic ketones demonstrated that counterion and solvent effects could reverse the above migratory preference. The DFT calculations provided a rationale for this 1,2-migration regiodivergency.

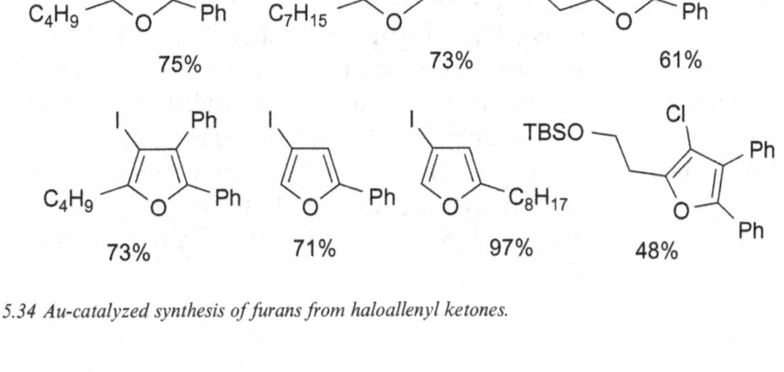

Scheme 5.34 Au-catalyzed synthesis of furans from haloallenyl ketones.

Scheme 5.34

In a glovebox under nitrogen atmosphere, to a 3.0 ml Wheaton micro-reactor equipped with a spin vane and screw cap with a PTFE-faced sili-cone septum under nitrogen atmosphere was added 6.2 mg (0.02 mmol, 2 mol%) of AuCl$_3$. The microreactor was removed from the glovebox and 1 ml of anhydrous toluene and 360 mg (1.02 mmol) of 4-bromo-1,2-diphenyl-2,3-octadien-1-one were sequentially added and stirred for 24 h. After the reaction was judged compete by TLC, it was quenched by filtra-tion through a pad of alumina with dichloromethane, concentrated, and purified over 30 ml of silica gel using hexanes as eluent to afford 270 mg (0.75 mmol, 75%) of 3-bromo-2-butyl-4,5-diphenyl furan as a colorless oil which solidifies upon storage.

Later on, they studied the [1,2]-alkyl shift in allenyl ketones [61]. Fully carbon-substituted and fused furans can be prepared effectively by this metal-catalyzed method (Scheme 5.35). In the optimization process, they found that Au^I and Au^{III} halides gave low yields of the furan, while nearly quantitative yield can be achieved with cationic Au^I complexes. Pt^{II}, Pt^{IV}, and Pd^{II} salts were inefficient in this reaction. Use of Cu^I halides resulted in no reaction, while employment of cationic Ag^I, Cu^I, and Cu^{II} salts produced the furan in moderate to high yields. They also tested Al, Si, Sn, and In triflates in this reaction, and moderate to excellent yields can be achieved. Then they chose In(OTf)$_3$ and Sn(OTf)$_2$ as the catalysts for the substrates testing.

Scheme 5.35

To an oven-dried ChemGlass pressure tube (or 5 ml Wheaton vial) charged with Sn(OTf)$_2$ (8.3 mg, 0.02 mmol, 5 mol%) as the catalyst and 3.6 ml of anhydrous toluene was added 1,4,4-triphenylbuta-2,3-dien-1-one (118.4 mg, 0.4 mmol) in 0.4 ml of anhydrous toluene under argon atmosphere and the reaction mixture was stirred at 100°C for 6 h. The reaction mixture was allowed to cool to room temperature and triethylamine (0.017 ml, 0.12 mmol) was added to quench the catalyst. The reaction mixture was filtered through a layer of Silica (EtOAc—eluent), the solvents were removed, and the residue was purified via flash Silica column chromatography (1:10:0.08 benzene/hexanes/Et$_3$N) to afford 96.3 mg (0.325 mmol, 81%) of 2,3,5-triphenylfuran as white crystalline solid.

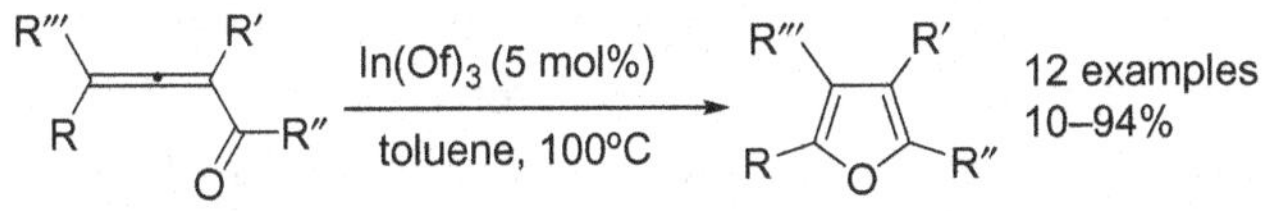

Scheme 5.35 In- or Sn-catalyzed synthesis of furans from allenyl ketones.

Marshall and co-workers explored the application of silver salt as catalyst in the synthesis of furans [62,63]. 2,3,5-Trisubstituted furans can be obtained by the cyclization of allenyl ketones and aldehydes upon treatment with $AgNO_3$-$CaCO_3$, in aqueous acetone. Later on, they succeeded to apply this methodology in the total synthesis of the (±)-Kallolide B (Scheme 5.36) [64]. In 2005, this methodology was applied in the synthesis of enantiomerically pure 2-(10-aminomethyl) furan derivatives [65].

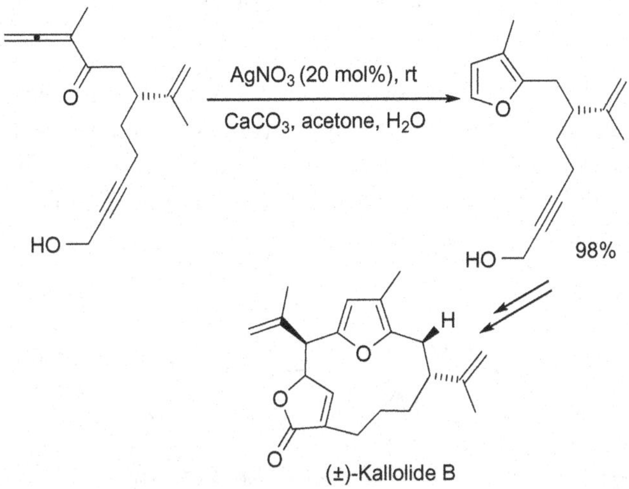

Scheme 5.36 Total synthesis of (±)-Kallolide B.

Hashmi's group studied the synthesis of allenyl ketones and their palladium-catalyzed cycloisomerization/dimerization in 1999 [66]. With $PdCl_2(MeCN)_2$ as the catalyst, 2-substituted furans and the 2,4-disubstituted furans can be formed. The yields and ratios of these products strongly depended on the nature of the groups being present. With the aryl thioether and the γ-halogen allenyl ketones the palladium-catalyzed reaction failed. Meanwhile, they studied mercury (II), silver(I), palladium(II), and bronsted acid as catalysts for this reaction as well [67]. The allenyl p-methoxybenzyl ketone and allenyl p-siloxybenzyl ketones selectively delivered three different products with three different transition metal-catalysts. With Hg(II)-catalysts a spiro[4.5]decene, with Ag(I)-catalysts a 2-substituted furan and with Pd(II)-catalysts, a 2,4-disubstituted furan was formed. Only with perchloric acid the intermolecular addition of water to the allene, leading to 1,3-dicarbonyl compounds, was observed. While with the corresponding allenyl o-methoxybenzyl ketone the Ag(I)- and Pd(II)-catalysts provided the expected products, the mercury-catalyst led to a new and interesting side-product which combined both the furan moiety and the spiro[4.5]decene moiety. Efforts to prepare allenyl hydroxybenzyl ketones failed; in one case a small amount of a 5H-benzo[b]oxepin-4-one was isolated. It was also not possible to extend the spirocyclization to allenyl p-siloxyphenyl ketone or allenyl 2-(p-siloxyphenyl)ethyl ketone. More recently, palladium catalyst was applied in the cyclization of 2-(1-alkynyl)-2-alken-1-ones as well [68].

In 2014, Miao et al. reported their work on CuI-mediated electrophilic cyclization reaction of cyclopropylideneallenyl ketones [69]. Highly substituted furan derivatives were prepared in good to excellent yields in the presence of I_2/CuI under mild conditions (Scheme 5.37).

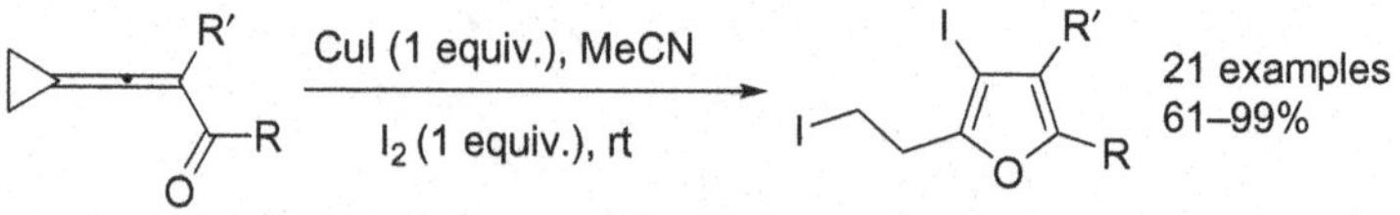

Scheme 5.37 Cu-mediated synthesis of furans from cyclopropylideneallenyl ketones.

In 2003, Gevorgyan and co-workers developed a novel copper-catalyzed 1,2-migration of the thio group in thioallenyl ketones and thioallenyl imines [70]. 3-Thio-substituted furans and pyrroles were effectively prepared under the assistant of CuI (Scheme 5.38). Recently, Oh and co-workers studied the synthesis of allenyl ketones and propargyl ketones from defined β-halovinyl ketones in the presence of NEt$_3$ [71]. Then under the assistant of CuCl (1 mol%), the *in situ* formed allenyl ketones and propargyl ketones were transformed into the corresponding 2,5-disubstituted furans.

Scheme 5.38

A mixture of propargyl ketone (246 mg, 1.0 mmol), CuI (12 mg, 0.05 mmol), and anhydrous DMAc (2.0 mL) was stirred in a Wheaton microreactor (3 mL) under an Ar atmosphere at 130°C. The reaction was monitored by TLC and GC/Ms until completion. After 12 h, the mixture was cooled to room temperature and poured into saturated aqueous NH$_4$Cl (20 ml). The phases were separated, and the aqueous phase was extracted (hexanes, 2 × 10 ml). The combined organic extracts were washed (brine, 10 ml), dried (Na$_2$SO$_4$, 2 g), and concentrated under reduced pressure. The residue was purified by means of silica-gel chromatography with hexanes to give pure furan.

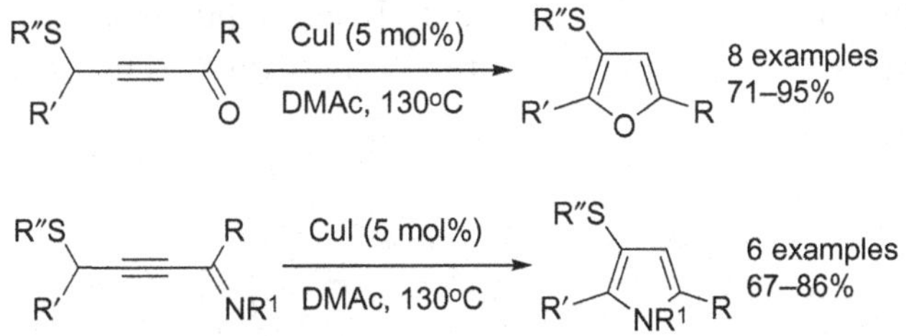

Scheme 5.38 Cu-catalyzed synthesis of 3-thio-substituted furans.

One year later, they reported the 1,2-migration of acyloxy, phosphatyloxy, and sulfonyloxy groups in allenes [72]. In the presence of silver catalyst, tri- and tetrasubstituted furans were synthesized in good yields (Scheme 5.39). Not only alkynyl ketones, but also its precursors can be applied as substrates. In their mechanistic studies, they found that the phosphatyloxy migration in conjugated alkynyl imines in their cycloisomerization to N-fused pyrroles proceeded via a [3,3]-sigmatropic rearrangement, but the analogous cycloisomerization of skipped alkynyl ketones proceeded through two consecutive 1,2-migrations, resulting in an apparent 1,3-shift, followed by a subsequent 1,2-migration through competitive oxirenium and dioxolenylium pathways. The mechanism of cycloisomerization of skipped alkynyl ketones containing an acyloxy group was found to be catalyst-dependent. Lewis and Brønsted acid catalysts caused an ionization/ S_N1' isomerization to the allene, followed by cycloisomerization to the furan, whereas transition metal catalysts evoked a Rautenstrauch-type mechanistic pathway [73].

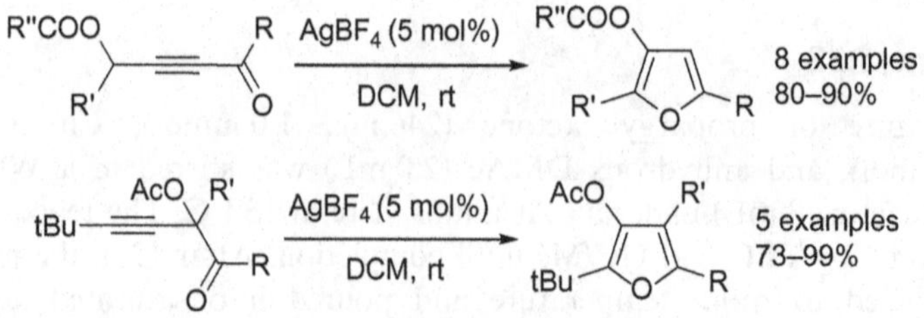

Scheme 5.39 Ag-catalyzed synthesis of acyloxy-substituted furans.

Wang and co-workers reported the 1,4-migration of the sulfanyl group in allenyl sulfides in 2007 [74]. This reaction can give multisubstituted furan products in good to excellent yields (Scheme 5.40). Additionally, in combination with a metal-catalyzed [2,3]-sigmatropic rearrangement, a one-pot sequential catalytic transformation of α-diazocarbonyl compounds to furan derivatives was realized as well.

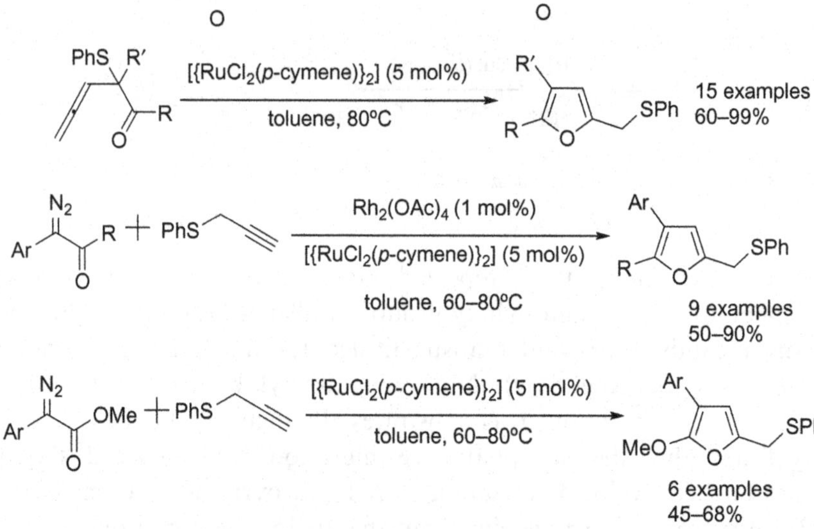

Scheme 5.40 Ru-catalyzed synthesis of sulfanyl-substituted furans.

An interesting method for synthesis of vinyl furans was developed by Liu's research group in 2011 [75]. Cycloisomerizations of the corresponding [3]cumulenones catalyzed by gold(I) complexes ((PPh₃)AuCl/AgOTf; 2 mol%) gave the desired furans in good yields through activation of the cumulenic double bond via a π-complex, and during the process, deprotonation from an alkyl group on cumulene terminus takes place to induce the isomerization. Additionally, synthesis of π-conjugated dienynes with high stereoselectivity was achieved by dehydration reaction of TMS-substituted [3]cumulenols catalyzed by TsOH · H$_2$O as well.

Ma and co-workers developed an efficient and highly selective procedure for the synthesis of 3,4-fused bicyclic furans [76]. The reaction proceeded through a rhodium(I)-catalyzed cycloisomerization of readily available 1,5-bis(1,2-allenylketone)s under mild conditions. Compared with [PdCl$_2$(MeCN)$_2$], [RhCl(CO)$_2$]$_2$ shows an excellent selectivity at room temperature (RT). This may be explained by the different bonding nature of palladium and rhodium to the carbon or oxygen in the organometallic intermediates. In addition, the selectivity of unsymmetric substrates has been nicely controlled by the electronic effect of nonallenyl groups connected to the carbonyl functionalities.

More recently, Zhu's group reported a palladium-catalyzed divergent cyclization, including cycloisomerization and aerobic oxidative cycloisomerization of homoallenyl amides [77]. Varieties of functionalized 2-amino-5-alkylfurans and 2-amino-5-formylfurans can be selectively synthesized in good to excellent yields (Scheme 5.41). Based on their mechanistic studies, they showed that peroxide may be a key intermediate for this Pd-catalyzed radical aerobic oxidative cycloisomerization of homoallenyl amides.

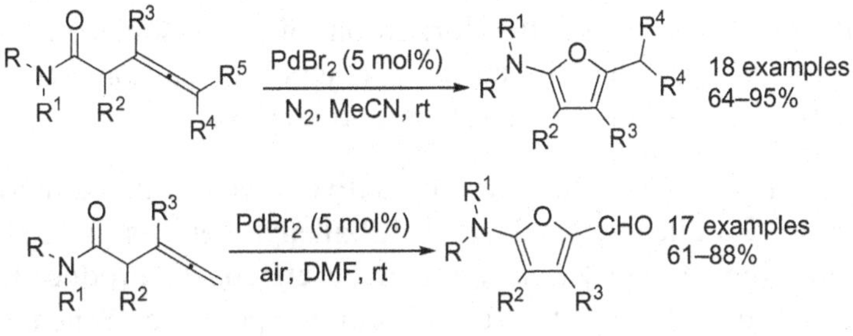

Scheme 5.41 Pd-catalyzed synthesis of 2-amino-substituted furans.

In 1995, Hashmi and co-workers reported his work on palladium-catalyzed dimerization of allenyl ketones [78]. Terminal allenyl ketones underwent cycloisomerization/dimerization under the assistant of palladium catalyst to give the corresponding furans in good yields (Scheme 5.42). Both tetrakis(2,2.2-trifluoroethoxycarbonyl)palladacyclopentadiene and Pd(OAc)$_2$ are effective for this reaction. Only furans from monocyclization of allenyl ketone were formed when $AgNO_3$, $[Rh_2(OAc)_4]$, $CuCl$, or $[RuCl_2(CO)_3]_2$ was applied as the catalyst. They performed mechanistic study as well [79]. With $PdCl_2(MeCN)_2$ as the catalyst in acetonitrile, the dimers could be obtained in good yields under mild conditions. Neither water nor oxygen needs to be excluded for a successful catalysis. Many functionalities and protecting groups that are frequently used in organic synthesis were tolerated. They also applied this concept in the synthesis of 2,4-furanophanes by palladium-catalyzed macrocyclization reactions of 1,n-diallenyl diketones [80].

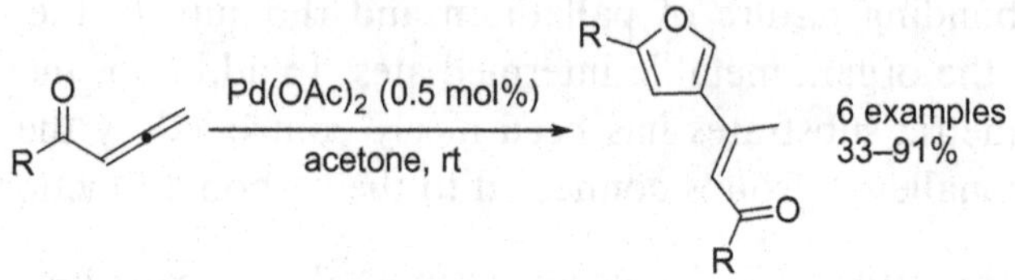

Scheme 5.42 Pd-catalyzed cycloisomerization/dimerization of allenyl ketones.

Alcaide et al. further studied this type of dimerization in 2007 [81]. The role of the steric interactions in this Pd^{II}-catalyzed transformation of terminal α-allenones was explored. By using $[PdCl_2(MeCN)_2]$ as the catalyst, the cycloisomerization/dimerization ratio of α-allenones is controlled by the substitution of the allene compound: unsubstituted allenones mainly afford dimerization, whereas allenones bearing an internal substituent favor the formation of cycloisomerization products. They showed that the mode of reaction (cycloisomerization vs. dimerization) of α-allenones is substrate-directable.

Ma's group reported the oxidative cyclization—dimerization reaction of 2,3-allenoic acids and 1,2-allenyl ketones in 2002 [82]. (3′-Furanyl)butenolide derivatives can be effectively produced in the presence of palladium catalyst at room temperature (Scheme 5.43). They found that highly optically active butenolides could also be easily formed from the optically active 2,3-allenoic acids, which was obtained conveniently through chiral resolution with optically active amines, that is, cinchonidine or α-methyl benzylamine [83]. In their mechanistic study, they showed that the reaction proceeded via a matched double oxypalladation-reductive elimination process. The Pd^{II} species may be regenerated via the subsequent cyclometallation of two equivalents of 1,2-allenyl ketones with Pd^0 and protonlysis of Pd enolates formed with the in situ generated HCl.

Scheme 5.43

[PdCl$_2$(MeCN)$_2$] (3 mg, 5 mol%) was added to a mixture of 2,3-allenoic acid (0.248 mmol) and 1,2-alkenyl ketone (1.26 mmol) in CH$_3$CN (3 ml), and the mixture was stirred at room temperature for 4 h. Evaporation and chromatographyon silica gel (petroleum ether/diethyl ether 3:1) afforded pure product.

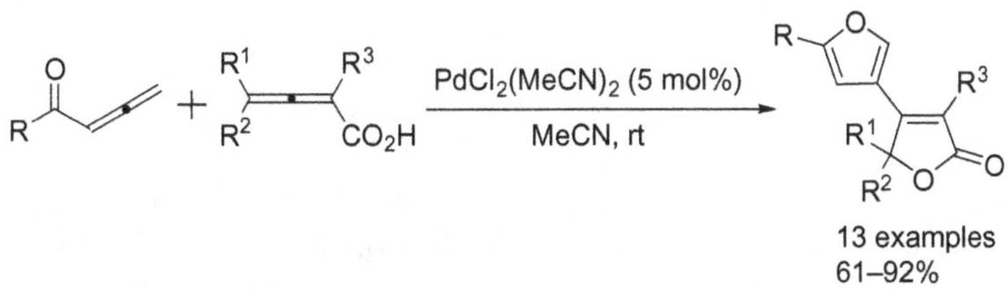

Scheme 5.43 Pd-catalyzed synthesis of (3′-furanyl) butenolides.

The use of 2,3-allenamides as coupling partners was explored as well [84]. In this case, 4-(furan-3′-yl)-2(5H)-furanimines were prepared by this Pd(II)-catalyzed oxidative heterodimerization reaction of with 1,2-allenyl ketones (Scheme 5.44). Benzoquinone was used as the oxidant here and reactions were performed in acetic acid.

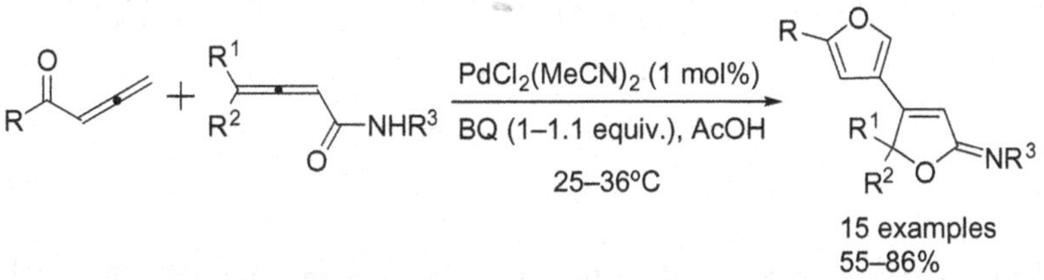

Scheme 5.44 Pd-catalyzed synthesis of 4-(furan-3′-yl)-2(5H)-furanimines.

Ma and co-workers studied the cyclization reaction of aryl or alk-1-enyl halides with 1,2-dienyl ketones as well [85,86]. Polysubstituted furans were produced in good to excellent yields from 1,2-allenyl ketones and organic halides with palladium as the catalyst in the presence of Ag$_2$CO$_3$ in toluene using Et$_3$N as the base (Scheme 5.45a). Allylic bromides can be applied as substrates as well [87]. 3-Allylic polysubstituted furans were formed from allylic bromides and 1,2-dienyl ketones via the reaction of a furanyl palladium intermediate with allylic bromide (Scheme 5.45b).

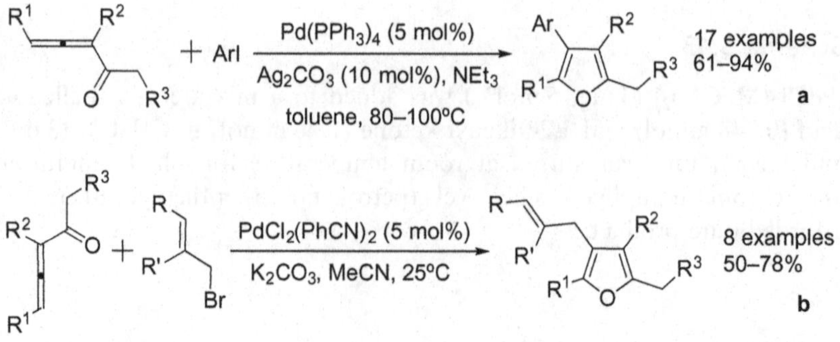

Scheme 5.45 Pd-catalyzed synthesis of furans from allenyl ketones and organic halides.

Recently, a palladium-catalyzed carbonylative dimerization of allenyl ketones was reported [88]. In the presence of p-benzoquinone (1 equiv.) under a CO atmosphere (balloon), difuranylketones were formed in moderate to good yields (Scheme 5.46). Mechanistically, the electron-withdrawing nature of the acyl group should enhance the electrophilicity of the acylpalladium species, and thus promote the oxypalladation of an additional molecule of allenyl ketones and leading to the difuranyl ketone formation.

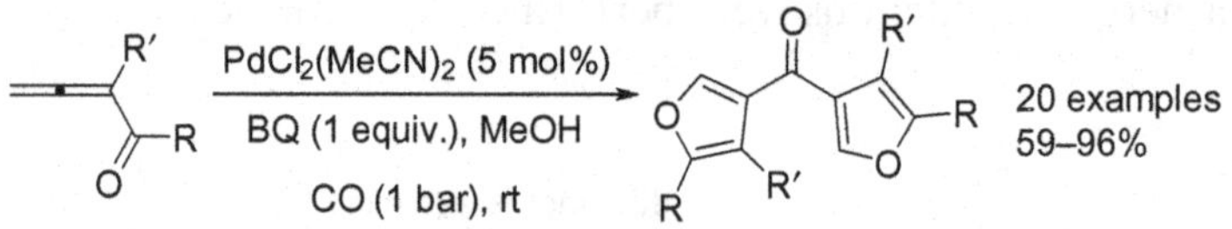

Scheme 5.46 Pd-catalyzed synthesis of difuranyl ketones.

Furans can be prepared from acetylenic ketones as well. As early as in 1988, Reisch and Bathe reported the use of CuI and NEt_3 as the catalyst system for cyclization of acetylenic ketones to the corresponding furans in low yields [89]. Gevorgyan's grop studied this copper-catalyzed transformation in 2002 [90]. 2-Monosubstituted and 2,5-disubstituted furans can be effectively produced from easily available alkynyl ketones in the presence of catalytic amounts of Cu(I). Furans possessing different functional groups, such as the sterically hindered t-Bu group, alkene moiety, alkoxy group directly attached to the furan ring, as well as a remote acid-sensitive Tetrahydropyranyl ether (OTHP) group, a base/nucleophile sensitive ester group, and an unprotected hydroxyl group, can all be prepared (Scheme 5.47).

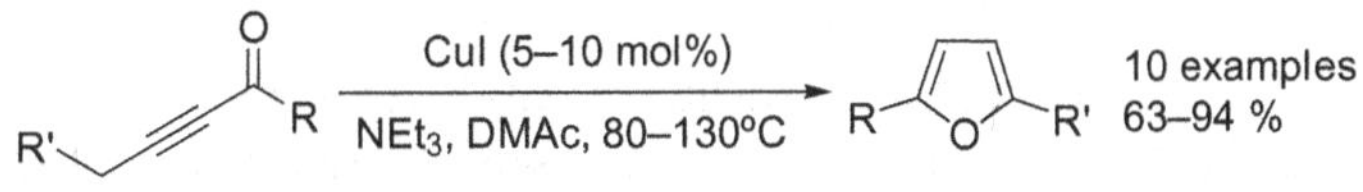

Scheme 5.47 Cu-catalyzed synthesis of furans from alkynones.

In 1986, Huang's group reported their study on palladium-catalyzed cyclization of alkynones to the corresponding furans [91]. In the presence of Pd(dba)$_2$ and PPh$_3$, good yields of the desired furans can be formed at 100°C. In 1991, Utimoto's group developed a palladium-catalyzed cyclization of acetylenic ketones to give the corresponding furans [92]. By intramolecular oxypalladation and subsequent protodemetalation, excellent yields of the corresponding furans can be produced (Scheme 5.48). 3-Allylfurans can be obtained as well by trapping the 3-furylpalladium species with allyl halides in the presence of 2,2-dimethyloxirane as a proton scavenger.

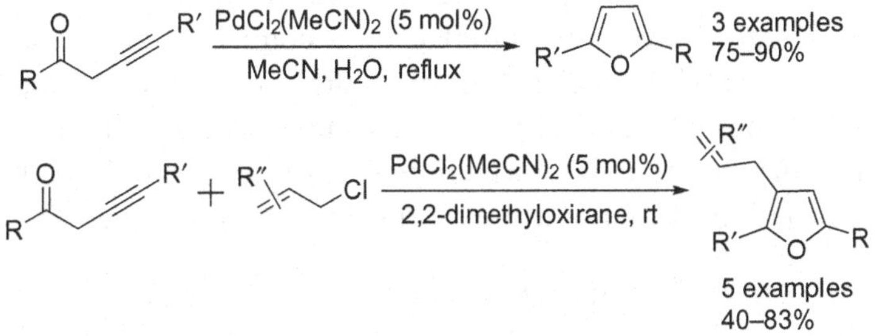

Scheme 5.48 Pd-catalyzed synthesis of furans from acetylenic ketones.

Ling and co-workers further studied palladium catalyst in the cyclization alkynones [93]. They found that alkynones undergo tandem dimerization and cyclization in the presence of PdCl$_2$(PPh$_3$)$_2$ and triethylamine in tetrahydrofuran at room temperature to give 3,3′-bifurans predominantly. Other palladium catalysts while under similar conditions, by rearrangement, lead to 2,5-disubstituted furans. They believe hydridopalladium halide intermediate from PdCl$_2$(PPh$_3$)$_2$ has been attributed to this selectivity. Furans and polysubstituted 3,3′-bifurans can be easily prepared from acetylenic ketones.

Later on, zinc-catalyzed cycloisomerizations of alkynones to furans were developed (Scheme 5.49) [94]. Through 5-*endo-dig* cycloisomerization of 1,4- and 1,2,4- mostly aryl-substituted but-3-yn-1-ones with catalytic amount of zinc chloride (10 mol%) as the catalyst in dichloromethane at room temperature, the desired 2,5-di- and 2,3,5-trisubstituted furans were formed in high yields (85–97%). Bicyclic furopyrimidine nucleosides can be prepared from acetyl-protected 5-alkynyl-2'-deoxyuridines by employing μ-oxo-tetranuclear zinc cluster $Zn_4(OCOCF_3)_6O$ as the catalyst. Furopyrimidine can be deprotected or simultaneously converted into pyrrolopyrimidine nucleoside.

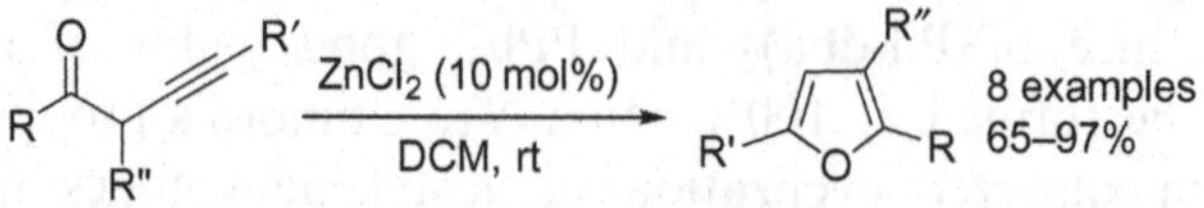

Scheme 5.49 Zn-catalyzed synthesis of furans from acetylenic ketones.

In 2004, Nishizawa and co-workers reported a mercuric triflate-catalyzed synthesis of furans [95]. 2-Methylfurans were produced from 1-alkyn-5-ones in excellent yields (Scheme 5.50). Benzene, toluene, and dichloromethane are all applicable solvents. For the reaction mechanism, the reaction should be initiated by π-complexation of an alkynyl group with $Hg(OTf)_2$ and forms oxonium cation after reacted with carbonyl group. Deprotonation of oxonium cation should provide vinyl mercury intermediate which will give the final 2-methylfuran after protonation, demercuration, and isomerization sequences.

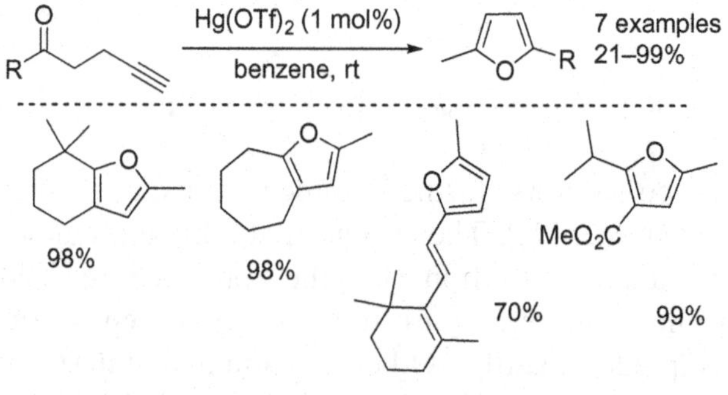

Scheme 5.50 Hg-catalyzed synthesis of 2-methylfurans.

An AgNO$_3$-catalyzed preparation of substituted furanopyrimidine nucleosides was reported in 2004 (Scheme 5.51) [96]. In the presence of AgNO$_3$ at room temperature, electrophilic cyclization occurred and provided the corresponding alkyl furanopyrimidine in quantitative yields (>95%). This procedure was also applied in the synthesis of fluorophores-based 5-phenylethynyluracils [97].

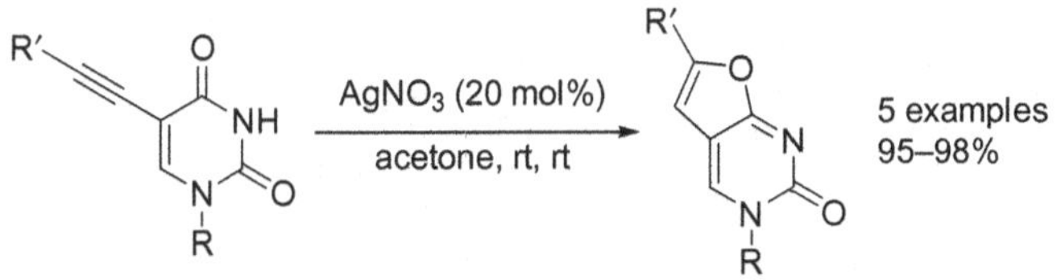

Scheme 5.51 Ag-catalyzed synthesis of furanopyrimidine.

Wipf and co-workers studied the cyclization of 3-aryl-substituted γ,δ-acetylenic ketones [98]. The substrates underwent a formal intramolecular S_N2'-*O*-alkylation with a variety of ether-type leaving groups to provide the corresponding 2,4,5-trisubstituted vinylfurans (Scheme 5.52). They succeeded in applying their procedure in the total synthesis of lophotoxin and pukalide as well with a stereoselective conversion of alkynoate to trimethylsilyl 2-alkenylfuran as the key step [99].

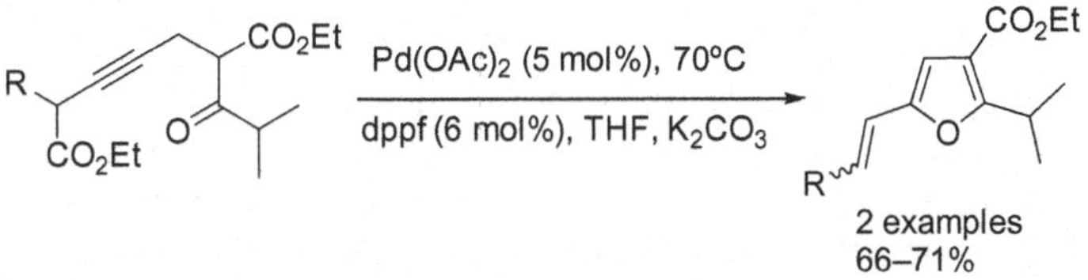

Scheme 5.52 Pd-catalyzed synthesis of 2,4,5-trisubstituted vinylfurans.

Belting and Krause studied gold catalyst in the cycloisomerization of alk-4-yn-1-ones and found that three efficient pathways are possible (Scheme 5.53) [100]. Substrates with a terminal or internal triple bond undergo a 5-*exo-dig* cycloisomerization to the corresponding multisubstituted furans under very mild conditions by using the cationic gold complex $Ph_3PAuOTf$ in toluene. Addition of catalytic amounts of *p*-TsOH·H_2O accelerates the reaction, but is not mandatory. A change in regioselectivity was observed with alkynones bearing two substituents at C-3 which will give 4*H*-pyrans by 6-*endo-dig* cycloisomerization. Finally, both substrate types afford alkylidene/ benzylidene-substituted tetrahydrofuranyl ethers when the reaction was carried out in an alcoholic solvent. These transformations probably proceed via enols or hemiacetals which act as internal nucleophiles for the attack at the triple bond which is activated by the gold catalyst.

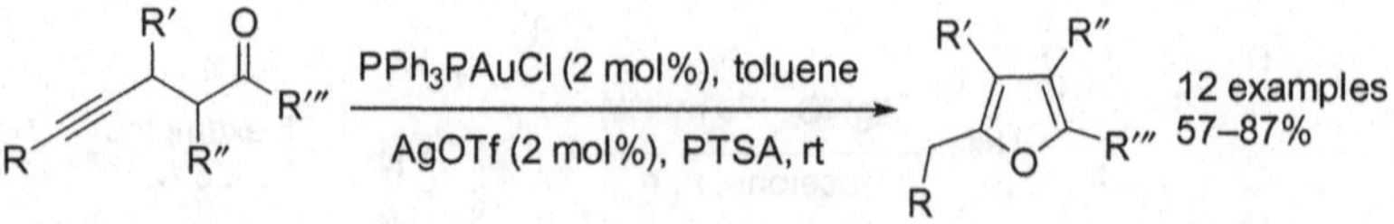

Scheme 5.53 Au-catalyzed cyclization of alk-4-yn-1-ones.

Zhang and co-workers reported a gold-catalyzed synthesis of 2-acylfurans from 3-(1-alkynyl)-2-alken-1-ones in 2010 (Scheme 5.54) [101]. This procedure represents a novel efficient and general methodology for the synthesis of 2-acylfurans with oxidation of the gold-carbene intermediates by H_2O_2 as one of the key steps. This method can be also applied in other gold-carbene intermediate involved reactions. Moreover this transformation was also reported with magnetic nano-supported Cu(I) catalyst ($Cu_2O@Fe_3O_4$) [102]. The catalyst can be easily recovered from the reaction by using external magnets and reused eight times without significant loss of its catalytic activity.

Scheme 5.54

To a solution of substrate (0.5 mmol) and $AuCl_3$ (0.025 mmol, 7.6 mg) in DCM (5 ml), H_2O_2 (30%, 1.5 mmol, 170 mg) was added. The resulting solution was stirred at rt until the reaction was completed (monitored by TLC). After the removal of solvent under reduced pressure, the residue was purified by column chromatography on silica gel (hexanes: AcOEt = 10:1) to give the desired product.

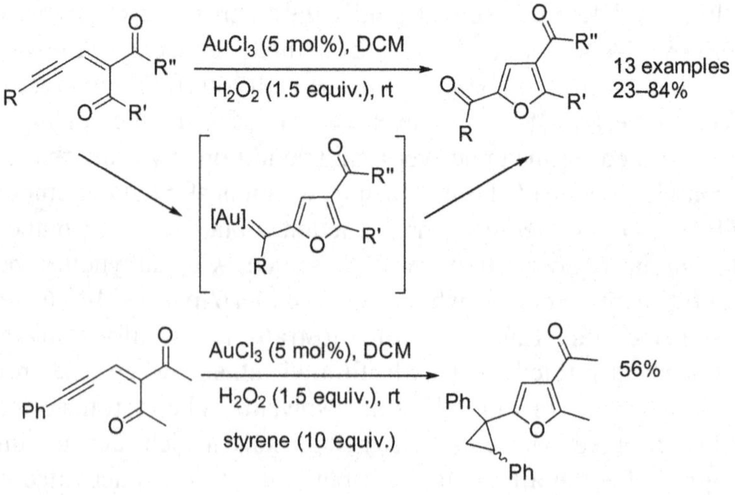

Scheme 5.54 Au-catalyzed synthesis of furans with Au-carbene as intermediate.

They developed a Pd(OAc)$_2$/CuI-co-catalyzed three-component reaction of 2-(1-alkynyl)-2-alken-1-ones, nucleophiles, and diaryliodonium salts as well [103]. Tetrasubstituted furans were produced in good to high yields under mild conditions (Scheme 5.55). A Pd$^{II/IV}$ catalytic cycle was proposed by the authors.

Scheme 5.55

A solution of yne-enone (0.3 mmol), Ph$_2$I$^+$PF$_6^-$ (255.6 mg, 0.6 mmol), Pd (OAc)$_2$ (3.4 mg, 0.015 mmol), and CuI (5.7 mg, 0.03 mmol) in DMSO (3 ml)/MeOH (0.3 ml) was stirred at 35°C under N$_2$. After completion of the reaction (monitored by TLC), water (5 ml) was added. The aqueous solution was extracted with diethyl ether (3 × 5 ml) and then the combined organic layer was washed with brine (20 ml). After drying (MgSO$_4$) and filtration, the organic layer was concentrated. The crude product was purified by column chromatography (petroleum ether:diethyl ether = 60:1) to give the pure product.

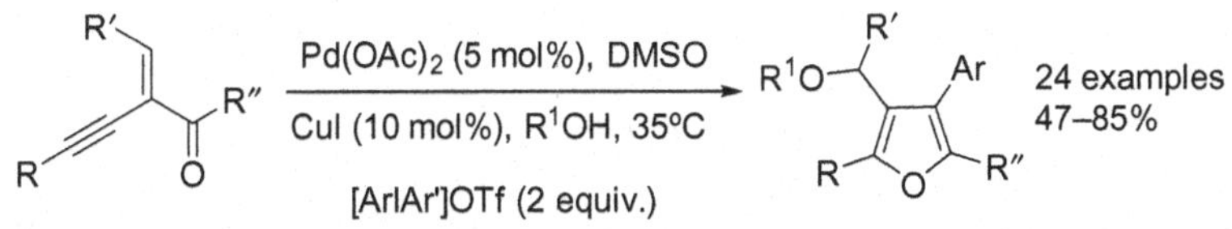

Scheme 5.55 Pd-catalyzed synthesis of tetrasubstituted 3-arylated furans.

Krafft and co-workers developed a tandem Au(III)-catalyzed heterocyclization/Nazarov cyclizations in 2011 [104]. Substituted carbocycle-fused furans were isolated in good yields (Scheme 5.56). An interesting dichotomy of reaction pathways as a function of solvent, confirmed by the isolation and trapping of reaction intermediates, and computational studies were performed as well, which supported the experimental findings.

Scheme 5.56

To a solution of 5 mol% of gold trichloride and 15 mol% of silver hexafluoro antimonate, under an argon atmosphere, a 0.2 M solution of ketone (0.2 mmol) in methylene chloride was added via syringe. The mixture was stirred for 3 h at ambient temperature before filtering through plug silica. The solution was then concentrated in vacuo and the product was isolated using silica gel column chromatography (ethyl acetate:hexane, 1:5 v/v).

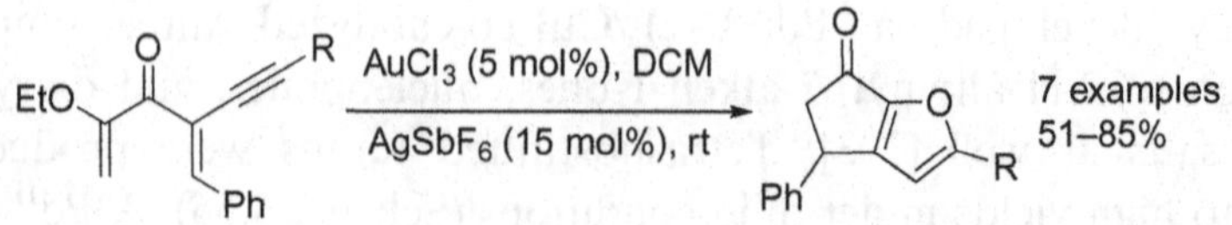

Scheme 5.56 *Au-catalyzed synthesis of furans via heterocyclization/Nazarov cyclizations.*

Kirsch's group developed two unprecedented domino reactions for the preparation of furan derivatives [105]. Started from 6-hydroxy-2-alkyl-2-alkynylcyclohexanones, furans can be prepared with $PtCl_4$ as the catalyst and gave 2,3-dihydrofurans in the presence of CuCl. For the furans, the reaction most likely proceeds through a heterocyclization followed by a ring-contracting 1,2-shift and a Grob-type fragmentation (Scheme 5.57).

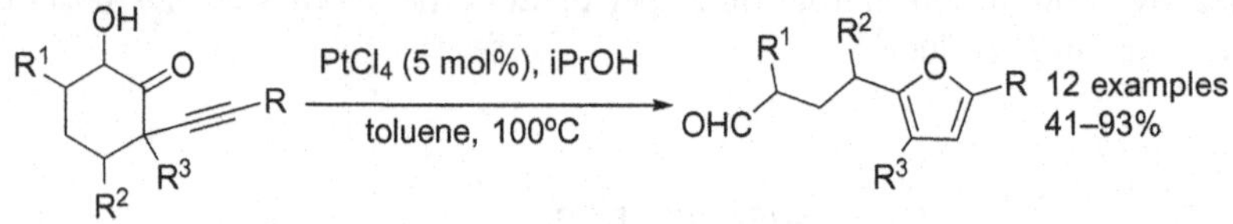

Scheme 5.57 *Pt-catalyzed synthesis of furans from alkynylcyclohexanones.*

Rao and co-workers reported the synthesis of di- and triarylfuran derivatives from but-2-ene-1,4-diones/but-2-yne-1,4-diones in 2003 [106]. In the presence of a catalytic amount of palladium on carbon using formic acid as reductant in poly(ethylene glycol)-200, furans were produced in good yields in a one-pot operation under microwave irradiation (Scheme 5.58). In some difficult cases, a catalytic quantity of concentrated sulfuric acid was required to promote the dehydrative-cyclization step.

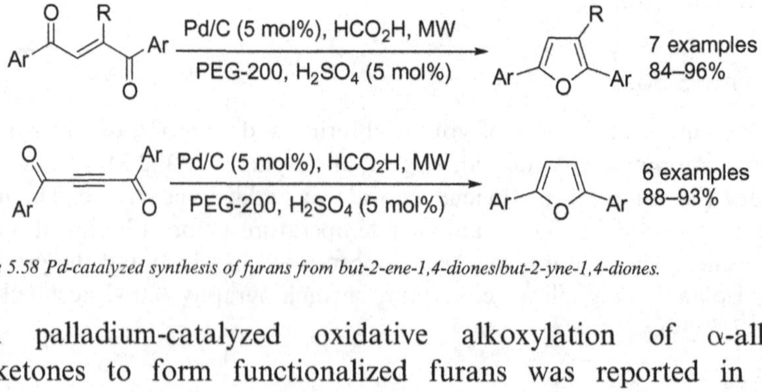

Scheme 5.58 *Pd-catalyzed synthesis of furans from but-2-ene-1,4-diones/but-2-yne-1,4-diones.*

A palladium-catalyzed oxidative alkoxylation of α-alkenyl β-diketones to form functionalized furans was reported in 2004

(Scheme 5.59) [107]. For example by treatment of 4-allyl-2,6-dimethyl-3,5-heptanedione with a catalytic amount of $PdCl_2(MeCN)_2$ (5 mol%) and a stoichiometric amount of $CuCl_2$ (2.2 equiv.) in 1,4-dioxane at 60°C for 12 h, the corresponding 3-isobutyryl-2-isopropyl-5-methylfuran can be produced in 77% isolated yield.

Scheme 5.59

A suspension of 4-allyl-2,6-dimethyl-3,5-heptanedione (0.51 mmol), $PdCl_2(MeCN)_2$ (7 mg, 0.026 mmol), and $CuCl_2$ (151 mg, 1.12 mmol) in 1,4-dioxane (5.1 ml) was stirred at 60°C for 12 h. The reaction mixture was cooled to room temperature, filtered through a plug of silica gel, and eluted with ether. The filtrate was concentrated under vacuum and chromatographed (hexanes-ether = 15:1) to give the pure product.

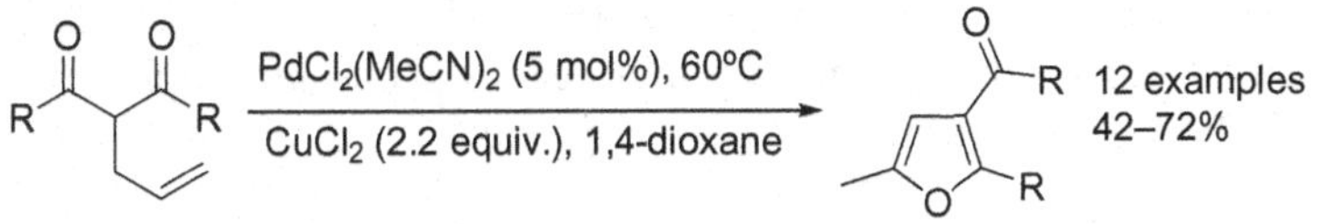

Scheme 5.59 Pd-catalyzed synthesis of furans from α-alkenyl β-diketones.

Dembinski's group developed a facile method for the synthesis of 2,5-disubstituted 3-bromo-4-fluoro- and 3-fluoro-4-iodofurans [108]. The furans were prepared in good yields by 5-*endo-dig* halocyclization of 2-fluoroalk-3-yn-1-ones with the use of *N*-iodo- and *N*-bromosuccinimide in the presence of gold chloride (5 mol%) and zinc bromide (20 mol%) in dichloromethane (Scheme 5.60). The 2-fluoroalk-3-yn-1-ones were prepared by monofluorination of the alk-1-en-3-yn-1-yl silyl ethers with Selectfluor which were prepared from 1,4-disubstituted alk-3-yn-1-ones.

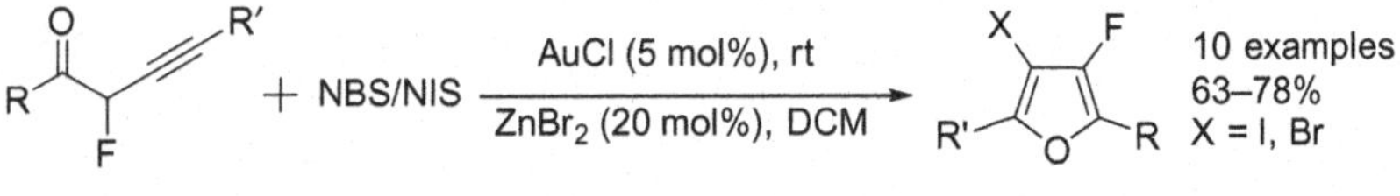

Scheme 5.60 Au-catalyzed synthesis of 3-fluorofurans.

In 2015, Li and co-workers reported a rhodium-catalyzed synthesis of multisubstituted furans from *N*-sulfonyl-1,2,3-triazoles [109]. The reaction involves an intramolecular trapping of α-imino carbene and subsequent aromatization. Good to excellent yields of the desired furans can be isolated (Scheme 5.61).

Scheme 5.61

A mixture of N-sulfonyl-1,2,3-triazoles (0.2 mmol), $Rh_2(esp)_2$ (1.5 mg, 1 mol%), and DCE (2 ml) was stirred under a nitrogen atmosphere at 100°C. After completion of the reaction (monitored by TLC), the resulting mixture was filtered rapidly through a funnel with a thin layer of silica gel and eluted with ethyl acetate. The filtrate was concentrated and the residue was purified by chromatography on silica gel (PE:EA = 5:1) to afford the desired products.

Or a mixture of N-sulfonyl-1,2,3-triazoles (0.2 mmol), $Rh_2(OAc)_4$ (0.9 mg, 1 mol%), and DCE (2 ml) was stirred under a nitrogen atmosphere at 100°C. After completion of the reaction (monitored by TLC), the resulting mixture was filtered rapidly through a funnel with a thin layer of silica gel and eluted with ethyl acetate. The filtrate was concentrated and the residue was purified by chromatography on silica gel (PE:EA = 4:1) to afford the desired products.

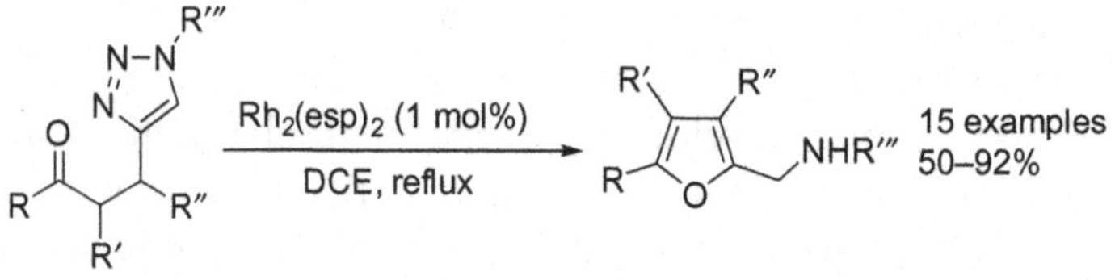

Scheme 5.61 Rh-catalyzed synthesis of furans from N-sulfonyl-1,2,3-triazoles.

Zhang and co-workers reported an enantioselective version of gold-catalyzed furan synthesis [110]. Furan-fused azepines derivatives can be prepared in good to excellent yields with high enantioselectivities by this redox-neutral domino reaction (Scheme 5.62).

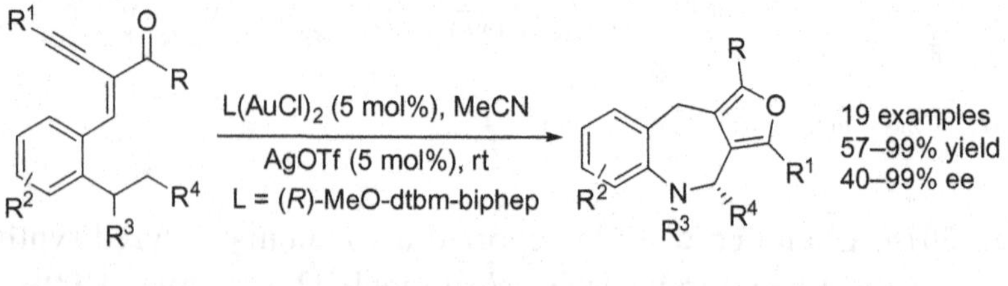

Scheme 5.62 Au-catalyzed synthesis of furan-fused azepines.

Mal et al. reported a gold(III) chloride-catalyzed synthesis of chiral-substituted 3-formyl furans from 3-benzyloxy-2-benzyloxy-methyl-5-phenylethynyl-2,3-dihydropyran-4-ones in 2014 [111]. Highly functionalized chiral 3-formyl furans were produced in moderate to excellent yields from suitably protected 5-(1-alkynyl)-2,3-dihydropyran-4-ones using $AuCl_3$ catalyst and water as the nucleophile (Scheme 5.63). The protected 5-(1-alkynyl)-2,3-dihydropyran-4-ones were prepared from the corresponding monosaccharides following oxidation, iodination, and Sonogashira-coupling sequences. Substituted furo[3,2-c] pyridine derivative can be produced as well by following two-step reaction sequences included a $TiBr_4$-catalyzed reaction of the 3-formyl furan derivatives resulted in the 1,5-dicarbonyl compound and then treatment with NH_4OAc under slightly acidic conditions to afford substituted furo[3,2-c]pyridine.

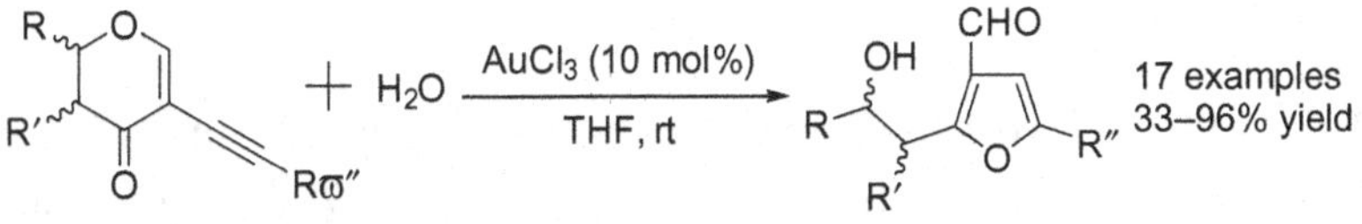

Scheme 5.63 Au-catalyzed synthesis of chiral substituted 3-formyl furans.

Oh and co-workers developed a Pt-catalyzed cyclization of enynones to produce highly substituted furans [112]. By Pt-catalyzed hydroxy- or alkoxy cyclization of 2-(1-alkynyl)-2-alkene-1-ones, a wide range of highly substituted furans were produced in good to excellent yields (Scheme 5.64). Meanwhile, Liang and co-workers studied this transformation with a catalytic amount of $Bu_4N[AuCl_4]$ in [bmim]BF_4 as well [113]. Good yields can be obtained at room temperature.

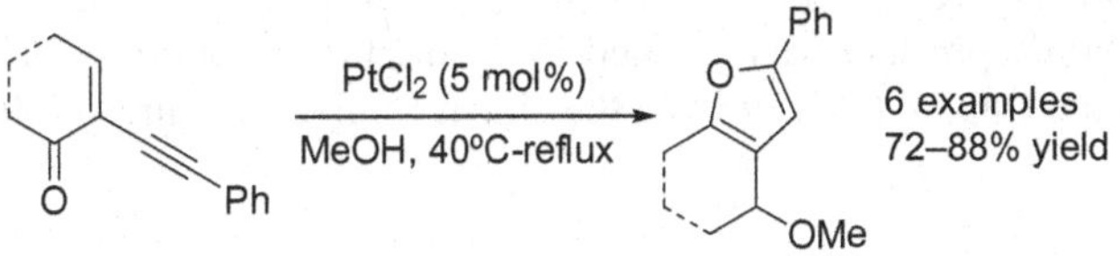

Scheme 5.64 Pt-catalyzed synthesis of substituted furans from enynones.

A copper-catalyzed synthesis of highly substituted furan from 3-(1-alkenyl)-2-alkene-1-al was established in 2010 [114]. By addition of water and followed by oxidation and unusual cyclization to naphthofuran ring, the desired furan derivatives can be produced in moderate yields (Scheme 5.65). In detail, the Cu(I) catalyst firstly coordinates with the triple bond to activate the carbonyl group to form oxonium complex. Then water molecule attacks the carbon atom adjacent to the oxygen atom to form another intermediate, which will give 1,4-dicarbonyl compound after reacted with oxygen molecule and rearrangement in the presence of water and also eliminate formic acid. This dicarbonyl compound forms naphthofuran ring with the elimination of water molecule.

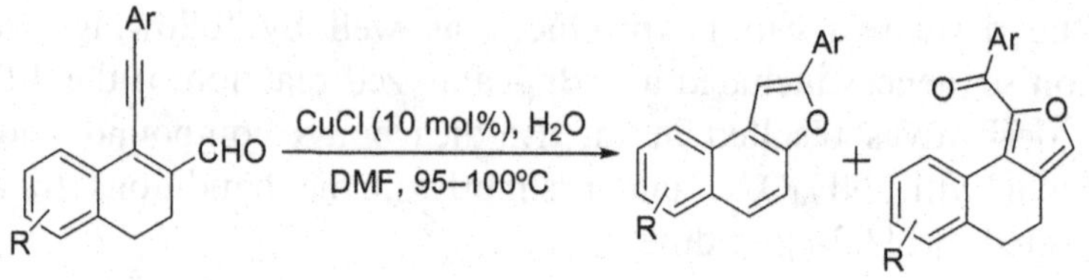

Scheme 5.65 Cu-catalyzed synthesis of naphthofurans.

Li and co-workers applied copper-catalyzed intramolecular *O*-vinylation of ketones with vinyl bromides in the synthesis of multisubstituted furans [115]. With CuI/3,4,7,8-tetramethyl-1,10-phenanthroline as the catalyst system, various ketones smoothly underwent the intramolecular *O*-vinylation with vinyl bromides to give the corresponding multisubstituted furans in moderate to excellent yields (Scheme 5.66).

Scheme 5.66

The mixture of ketone (0.30 mmol), CuI (6 mg, 0.030 mmol), 3,4,7,8-tetramethyl-1,10-phenanthroline (14 mg, 0.060 mmol), and Cs_2CO_3 (0.19 g, 0.60 mmol) in 1,4-dioxane (10 ml) was stirred at reflux for 20 h under nitrogen atmosphere. The resulting mixture was cooled down to room temperature and filtered. The filtrate was then concentrated in vacuo and the crude product was purified by flash chromatography on silica gel with hexane/EtOAc (10:1, v:v) as the eluent to give the pure product.

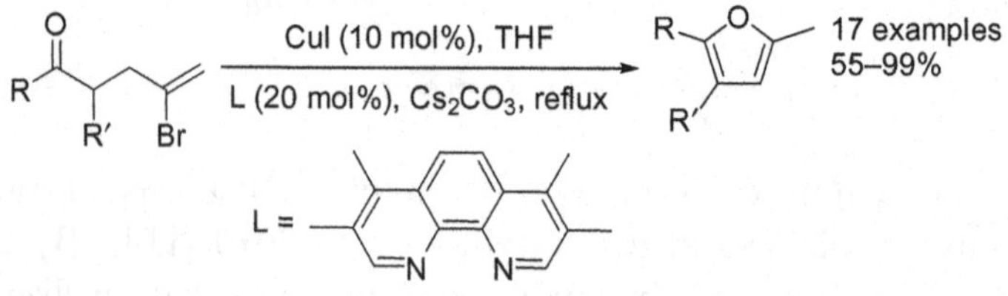

Scheme 5.66 Cu-catalyzed synthesis of furans from vinyl bromides.

AuCl₃ as a catalyst was also applied in the cycloisomerization of β-alkynyl β-ketoesters to the corresponding furans (Scheme 5.67a) [116]. The reaction went through alkyne activation by Au(III) and subsequent intramolecular attack by the ketone oxygen to form a zwitterionic intermediate. Then protodemetalation regenerates the gold catalyst and releases the heterocycle, which isomerizes to the furan product. More recently, gold catalyst was applied in the cycloisomerization of homopropargylic aldehydes and imines as well (Scheme 5.67b) [117]. Gold catalyst was also explored in the rearrangement of alkynyl sulfoxides [118]. Alkynyl sulfoxides were transformed into the corresponding benzothiepinones, benzothiopines, or α-thioenones via α-carbonyl gold-carbenoid intermediate. Interestingly, furan derivative can be produced if a 1,4-diyne moiety exists in the structure.

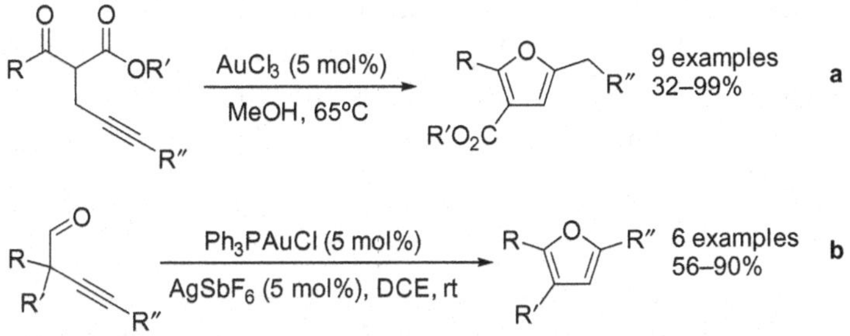

Scheme 5.67 Au-catalyzed ycloisomerization of homopropargylic aldehydes and ketones.

Barluenga and co-workers reported a Cu(I)-catalyzed regioselective synthesis of polysubstituted furans from propargylic esters in 2008 [119]. Proceeded via (2-furyl)carbene complexes as the intermediate, good yields of the desired furans can be achieved (Scheme 5.68).

Scheme 5.68

To a solution of the propargyl ester (0.5 mmol) and the corresponding silane (or germane) (1.5 mmol) in CH_2Cl_2 (10 ml) [Cu(CH₃CN)₄][BF₄] (7.9 mg, 0.025 mmol), was added. The mixture was stirred at room temperature until disappearance of the starting diyne (checked by TLC; 2–4 h). The solvent was distilled under reduced pressure and the residue was purified by flash chromatography (SiO₂, 10:1 mixture of hexanes and ethyl acetate) to give the corresponding furan derivatives.

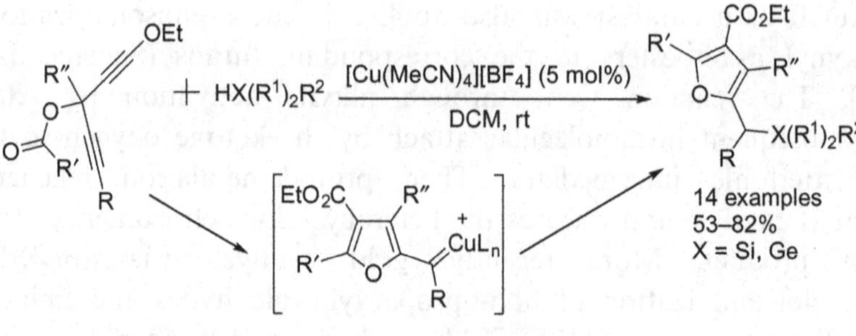

Scheme 5.68 Cu-catalyzed synthesis of furans from propargylic esters.

A palladium-catalyzed cycloisomerization of methylene- or alkylidenecyclopropyl ketones to the corresponding furans was reported in 2004 by Ma and co-workers [120]. The substrates, 2-methylene- or alkylidenecyclopropanyl ketones, were easily prepared by the regioselective cyclopropanation of allenes or the reaction of methylene-/alkylidenecyclopropanyllithium with N,N-dimethyl carboxylic acid amides. Good yields of the desired furans can be formed in the presence of $PdCl_2(CH_3CN)_2$ and sodium iodide (Scheme 5.69).

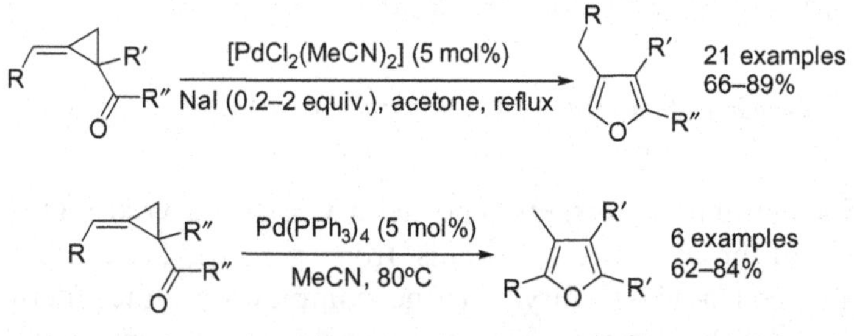

Scheme 5.69 Pd/NaI-catalyzed synthesis of furans.

Padwa and co-workers explored the application of rhodium catalyst in the cyclization of α-diazo alkynyl-substituted ketones [121]. Through the formation of vinyl rhodium carbenoids intermediates, furans can be effectively produced (Scheme 5.70a). Meanwhile, Hoye's group reported on the cyclization of α-diazo ketones as well [122]. Furans can be formed by using rhodium as the catalyst (Scheme 5.70b).

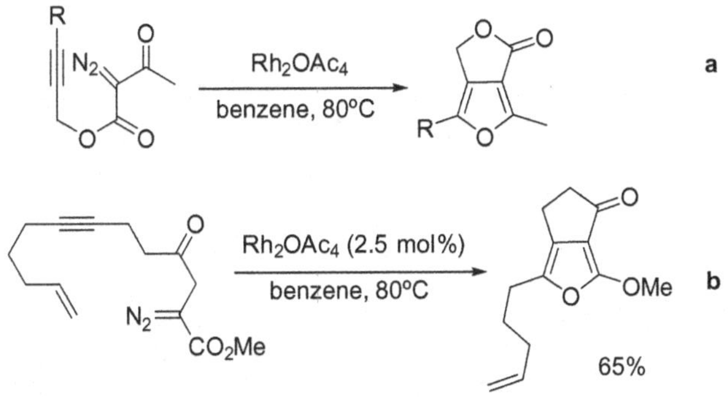

Scheme 5.70 Rh-catalyzed synthesis of furans from α-diazo ketones.

A gold-catalyzed cyclization of 2-alkynyl-1-cycloalkenecarbaldehydes was established by Oh and co-workers in 2008 [123]. Under the assistant of gold catalysts and through Au-carbene intermediates which formed by 5-*exo-dig* cyclization, which then reacted with a double bond to give the corresponding products in moderate yields (Scheme 5.71).

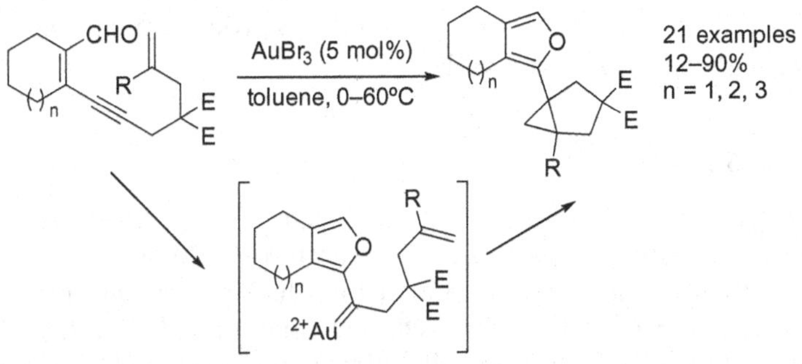

Scheme 5.71 Au-catalyzed synthesis of furans from 2-alkynyl-1-cycloalkenecarbaldehydes.

Zhang and Zhou developed a gold-catalyzed synthesis of ring-fused tetrahydroazepines in 2010 [124]. By using carbophilic gold(I) complex (IPrAuOTf) as the catalyst; moderate to excellent yields of the desired products can be obtained under the mild conditions (Scheme 5.72). Interestingly, using the same starting materials with oxophilic Sc(OTf)$_3$ as the catalyst, the reaction undergoes domino 1,5-hydride shift/cyclization to afford highly substituted multifunctionalized ring-fused tetrahydroquinolines in moderate to excellent yields with high diastereoselectivity.

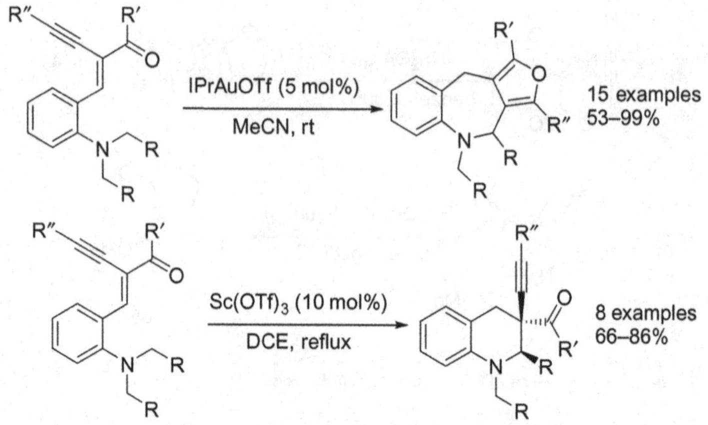

Scheme 5.72 Au-catalyzed synthesis of ring-fused tetrahydroazepines.

In 2015, Chang's group reported a Bi(OTf)$_3$-mediated procedure for the synthesis of furans [125]. By cycloisomerization of α-substituted γ-alkynyl arylketones, a series of 2-arylfurans were produced in good yields (Scheme 5.73). The structures of the key products were confirmed by X-ray crystallography and the procedure can be enlarged to gram scale without any problem.

Scheme 5.73

Bi(OTf)$_3$ (13 mg, 0.02 mmol) and molecular sieve (4 Å, 100 mg) were added to a solution of substrate (1.0 mmol) in dry MeNO$_2$ (5 ml) at rt. The reaction mixture was stirred at rt for 3 h. The reaction mixture was diluted with CH$_2$Cl$_2$ (10 ml), filtered, and concentrated. The residue was diluted with water (10 ml) and the mixture was extracted with CH$_2$Cl$_2$ (3 × 20 ml). The combined organic layers were washed with brine, dried, filtered, and evaporated to afford crude product under reduced pressure. Purification on silica gel (hexanes/EtOAc = 10/1 ∼ 4/1) afforded the pure product.

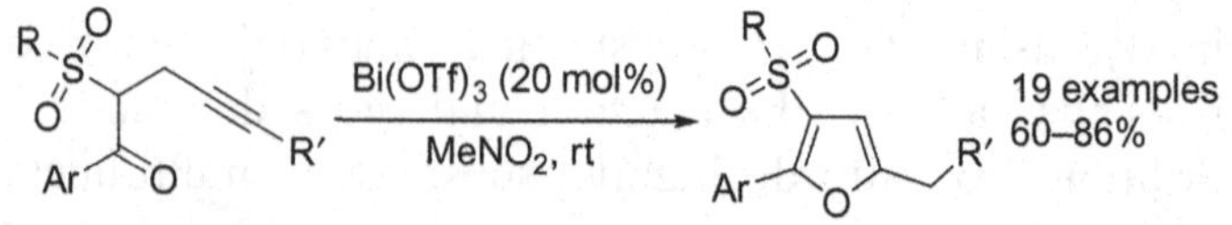

Scheme 5.73 Bi-mediated synthesis of furans from γ-alkynyl arylketones.

A rhodium-catalyzed polymerization of enyne ketones to the corresponding furylcyclopropane-containing polymers and furfurylidene-containing polymers was reported in 2004 [126]. Through the *in situ* generation of (2-furyl)carbene complexes by reacted [Rh(OAc)₂]₂ catalyst with substrates, moderate to good yield can be achieved (Scheme 5.74).

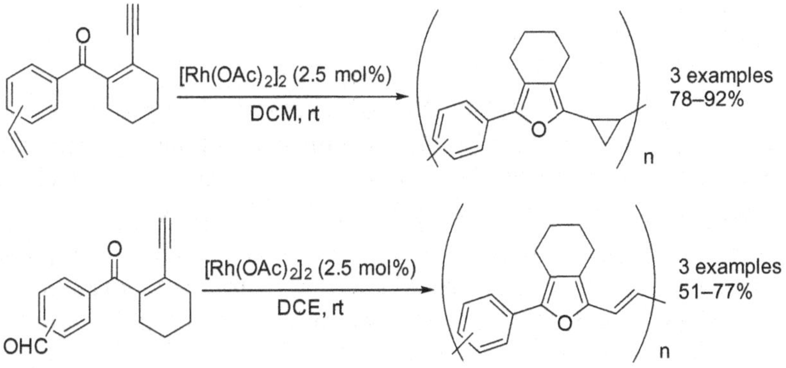

Scheme 5.74 Rh-catalyzed synthesis of furan-containing polymers.

Ma and Zhang reported a palladium(II)-catalyzed regioselective cycloisomerization of alkylidene cyclopropyl ketones in 2003 (Scheme 5.75) [127]. Started from the same substrates, furans were prepared in the presence of two equivalents of sodium iodide or other salts through distal-bond cleavage; while polysubstituted 4*H*-pyrans were formed in the absence of salt by proximal-bond cleavage.

Scheme 5.75

[PdCl₂(MeCN)₂] (17 mg, 0.65 mmol) was added to a solution of substrate (1.3 mmol) and sodium iodide (400 mg, 2.7 mmol) in acetone (5 ml). The mixture was then stirred under reflux for 14 h. Evaporation of the solvent and chromatography on silica gel (petroleum ether/ether 100:1) afforded pure furan derivatives.

[PdCl₂(MeCN)₂] (6 mg, 5 mol%) was added to a solution of substrate (0.50 mmol) in acetone (2 ml). The mixture was then stirred at rt for 15 min. Evaporation of the solvent and chromatography on silica gel (petroleum ether/ether 100:1) afforded pure pyrans.

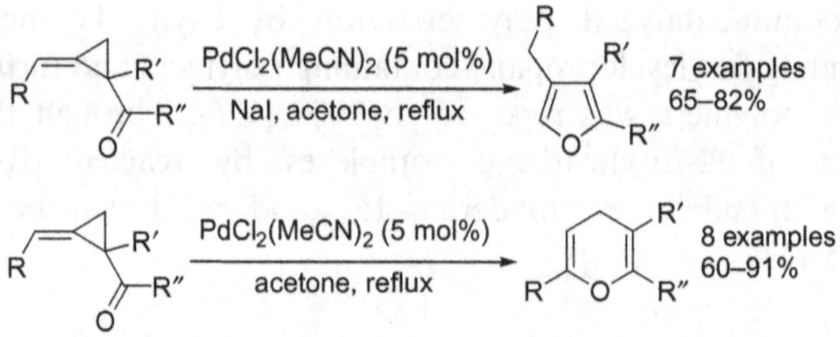

Scheme 5.75 Pd-catalyzed synthesis of furans and 4H-pyrans.

In 2006, Schmalz and Zhang reported a gold(I)-catalyzed reaction of 1-(1-alkynyl)-cyclopropyl ketones with nucleophiles [128]. Highly substituted furans were produced in good yields under mild conditions (Scheme 5.76). The substrates applied can be easily prepared from the corresponding enones through cyclopropanation. Two possible reaction pathways were proposed, started with the coordination of gold-catalyst to the triple bond or coordinate to both carbonyl and alkyne to give the chelate complex.

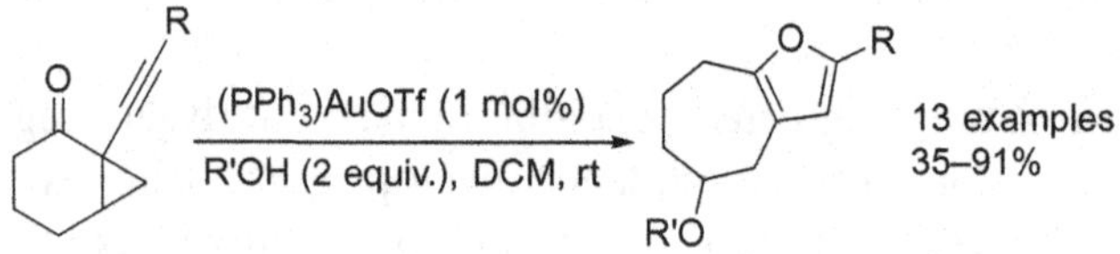

Scheme 5.76 Au-catalyzed synthesis of furans from 1-(1-alkynyl)-cyclopropyl ketones.

In 2008, Zhang and Xiao developed a [PdCl$_2$(CH$_3$CN)$_2$]-catalyzed three-component reaction for the synthesis of tetrasubstituted furans [129]. The reaction followed Michael addition/cyclization/cross-coupling reaction sequences. In this procedure, [PdCl$_2$(CH$_3$CN)$_2$] has two roles: activation of triple bond and organo halides. Concerning the nucleophiles, various types of nucleophiles such as O-, N-, C-based nucleophiles and olefin-tethered O-, N-, C-based nucleophiles can all be applied successfully (Scheme 5.77) [130]. By mechanistic studies, two possible reaction pathways were proposed (cross-coupling reaction versus Heck reaction) and the same intermediate furanylpalladium species were observed. The reaction pathway is dependent on the property of the nucleophile and on the length of the tethered chain as well. When olefin-tethered O-based nucleophiles were used, only the cross-coupling reaction pathway was observed; in contrast, both reaction pathways were observed when olefintethered *C*-based nucleophiles were employed. The product ratio is dependent on the length of the tethered chain. Additionally, ring-closing metathesis (RCM) of the corresponding furans with C=C bonds provides an easy method for the preparation of functionalized oxygen-heterocycles-3,4-fused bicyclic furans. Aryl iodides can be applied instead of allylic chlorides as well [131].

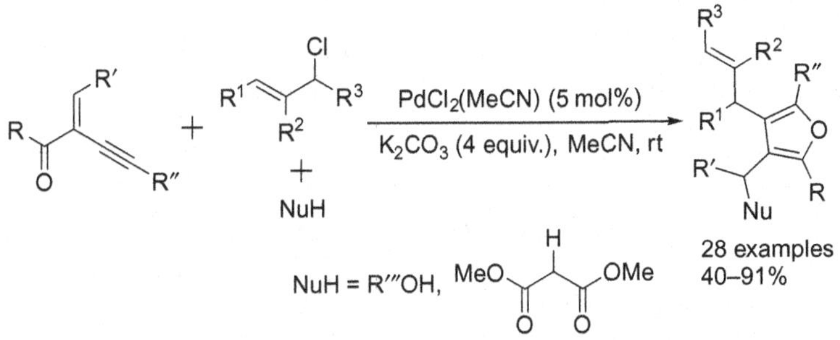

Scheme 5.77 Pd-catalyzed three-component synthesis of furans.

They studied the reaction of 2-(1-alkynyl)-2-alken-1-ones with nucleophiles and vinyl ketones as well [132]. Moderate to good yields of the corresponding furans can be produced under the same reaction conditions (Scheme 5.78). Interestingly, a double-addition reaction of two molecules of vinyl methyl ketones to the substrates can be observed when (*E*)-2-benzylidene-4-phenylbut-3-ynal was used as the substrate.

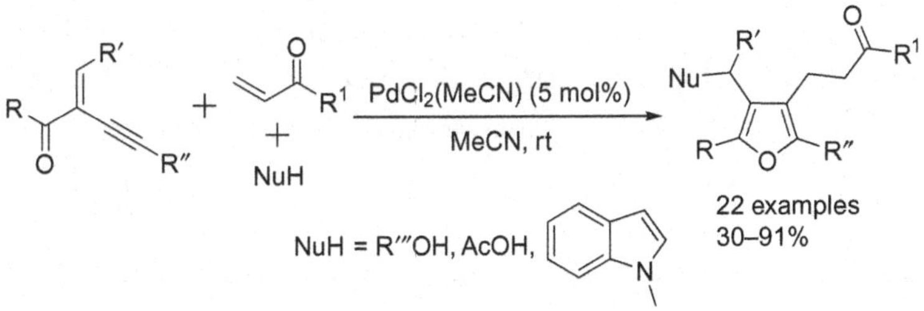

Scheme 5.78 Pd-catalyzed three-component synthesis of furans with vinyl ketones.

Zhang and co-workers studied the 1,3-dipolar cycloaddition of 2-(1-alkynyl)-2-alken-1-ones with nitrones in 2009 (Scheme 5.79a) [133]. In the presence of gold catalyst, highly substituted furo[3,4-*d*][1,2] oxazines can be produced in good yields with excellent selectivity (Scheme 5.79). By using (R)-MeO-biphep as the chiral ligand, moderate *ee* of the product can be observed. Notably, *ee* can be dramatically improved by using (R)-C$_1$-tunephos or (R)-MeO-dtmb-biphep as the ligand which were modified from (R)-MeO-biphep [134]. Recently, they reported a gold-catalyzed dehydrogenative heterocyclization of 2-(1-alkynyl)-2-alken-1-ones with electron-deficient olefins and alkynes as well (Scheme 5.79b) [135]. Good yields 2,3-furan-fused carbocycles can be produced in the presence of gold catalyst with pyridine *N*-oxide as the additive by trap the *in situ* furan-based *o*-QDMs with electron-deficient olefins and alkynes.

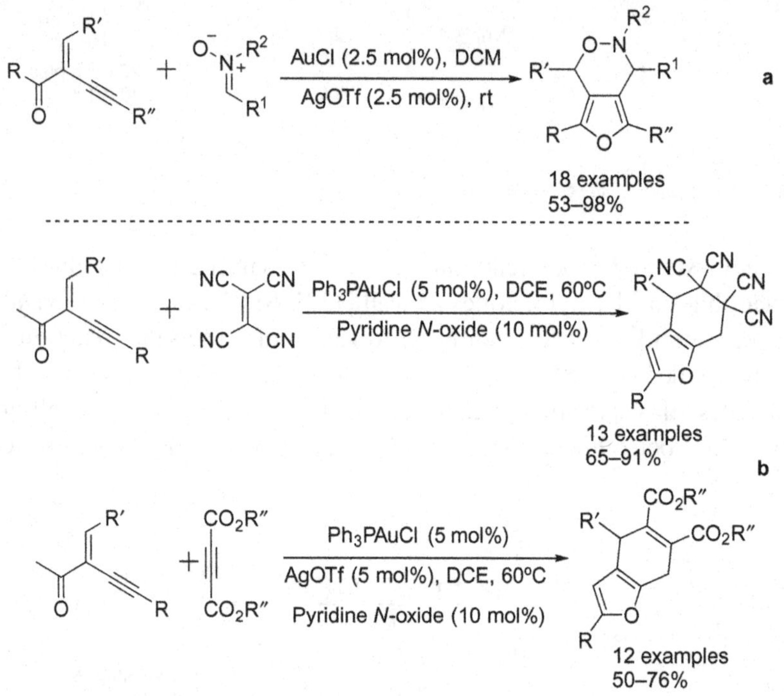

Scheme 5.79 Pd-catalyzed synthesis of furans from 2-(1-alkynyl)-2-alken-1-ones.

In the same period, Wang and co-workers developed a highly diastereoselective gold- or copper-catalyzed procedure for the synthesis of furans from 1-(1-alkynyl) cyclopropyl ketones and nitrones (Scheme 5.80) [136]. DFT calculation was carried out as well in order to understand the reaction mechanism.

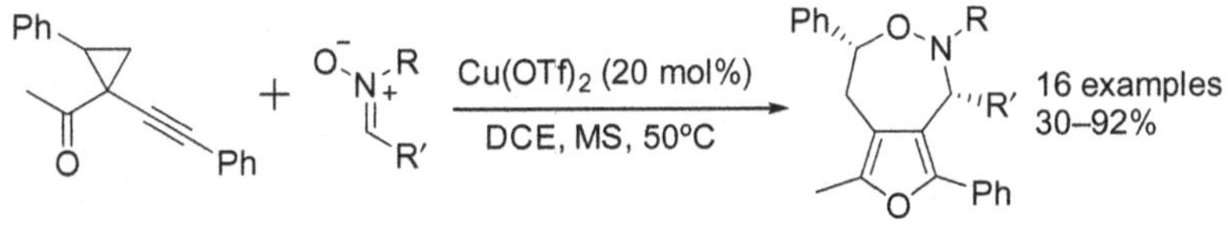

Scheme 5.80 Cu-promoted synthesis of furans from 1-(1-alkynyl) cyclopropyl ketones.

In 2004, Larock and co-workers reported their achievements on the gold-catalyzed (AuCl₃) cyclization of 2-(1-alkynyl)-2-alken-1-ones with various nucleophiles under mild reaction conditions [137]. In their studies, they found that the cyclization of 2-(1-alkynyl)-2-alken-1-ones can be induced by electrophile as well [138]. Highly substituted furans were formed in good to excellent yields under very mild reaction conditions (Scheme 5.81). Various nucleophiles, including functionally substituted alcohols, H₂O, carboxylic acids, 1,3-diketones, and electron-rich arenes, and a range of cyclic and acyclic 2-(1-alkynyl)-2-alken-1-ones, readily participate in these cyclizations. Iodine, NIS, and PhSeCl have proven successful as electrophiles in this process. The resulting iodine-containing furans can be readily elaborated to more complex products using known organopalladium chemistry. This transformation was reported with silver catalyst (AgSbF₆ (10 mol%), NaHCO₃ (1 equiv.), DCM, 0°C.) [139] and copper catalyst (CuBr (10 mol%), DMF, 80°C) as well [140].

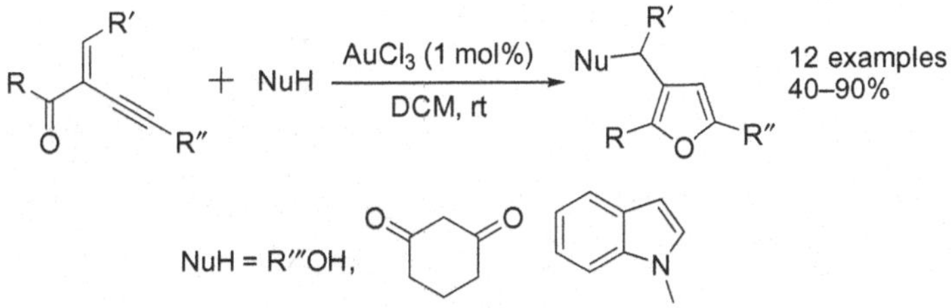

Scheme 5.81 Au-catalyzed synthesis of substituted furans with nucleophiles.

Zhang's group reported a rhodium-catalyzed domino heterocyclization and formal [(3 + 2) + 2] carbocyclization reaction for the construction of highly substituted furan derivatives (Scheme 5.82a) [141]. Started from diyne-enone and alkyne, fused tricyclic tricycloheptadienes were formed in low to good yields and the regioselectivity of this carbocyclization heavily depends on both tether atom and the nature of the alkyne. In the absence of alkynes, 2,3-fused bicyclic furans can be produced selectivity

(Scheme 5.82b) [142]. Notably, by simply increasing the pressure of CO, a tandem heterocyclization and carbonylative $[(3+2)+1]$ cyclization reaction occurred [143]. From diyne-enone and carbon monoxide, polycyclic furan scaffold was isolated in good yields (Scheme 5.82c).

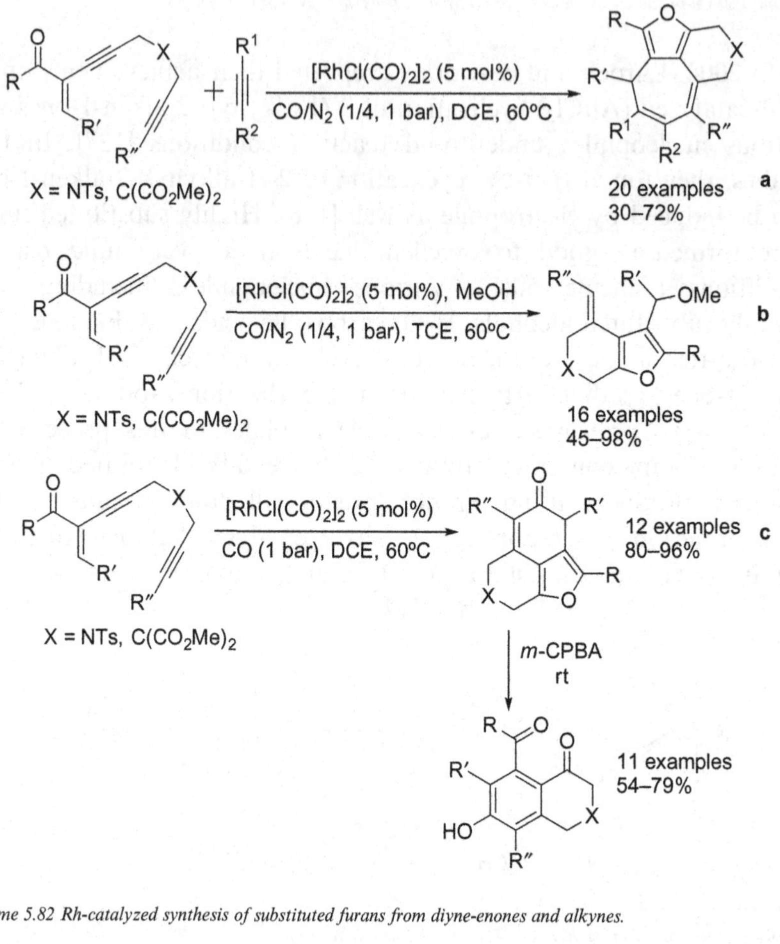

Scheme 5.82 Rh-catalyzed synthesis of substituted furans from diyne-enones and alkynes.

In the same year, they also explored gold catalyst in the synthesis of highly substituted furo[3,4-c]azepines (Scheme 5.83a) [144]. With 2-(1-alkynyl)-2-alken-1-ones and heterodienes as the substrates, furo [3,4-c]azepines were formed in good yields under mild conditions in stereoselective manner. The nature of the heterodienes will control the reaction pathway. Interestingly, with the same substrates, different products can be produced by using different gold catalyst [145].

Cyclopenta[c]furans can be efficiently constructed from readily available 2-(1-alkynyl)-2-alken-1-ones and 3-styrylindoles (Scheme 5.83b).

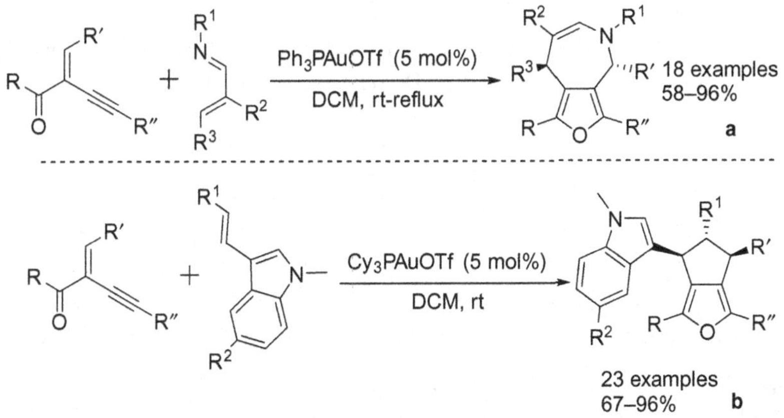

Scheme 5.83 Au-catalyzed synthesis of furo[3,4-c]azepines and cyclopenta[c]furans.

A palladium-catalyzed cycloisomerization-allylation of 4-alkynones to synthesis of 5-homoallylfuran derivatives was reported in 2011 [146]. With Pd₂(dba)₃ and ᵗBu₃P as the catalytic system and Cs₂CO₃ as the base, highly substituted furans having homoallyl groups at 5-position could be prepared in good yields with allyl carbonates in MeCN (Scheme 5.84).

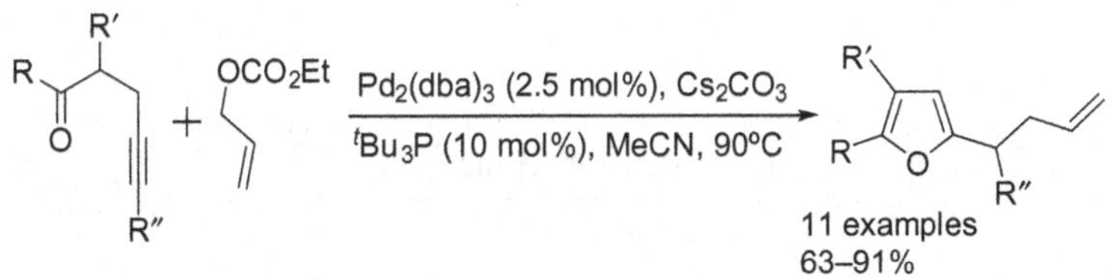

Scheme 5.84 Pd-catalyzed synthesis of 5-homoallylfuran derivatives.

A chromium-catalyzed synthesis of 5-phenyl-2-furylcyclopropane derivatives was established in 2002 [147]. In the presence of a catalytic amount of $Cr(CO)_5(THF)$, by reaction of alkenes with conjugated ene-yne-ketones, good yields of the desired furans can be isolated (Scheme 5.85). The key intermediate of this cyclopropanation is a (2-furyl)carbene complex generated by a nucleophilic attack of carbonyl oxygen to an internal alkyne carbon in π-alkyne complex or δ-vinyl cationic complex. Notably, other late transition metal compounds, such as $[RuCl_2(CO)_3]_2$, $[RhCl(cod)]_2$, $[Rh(OAc)_2]_2$, $PdCl_2$, and $PtCl_2$, can also catalyze this reaction effectively. When the reactions were carried out with dienes as a carbene acceptor, the more substituted or more electron-rich alkene moiety was selectively cyclopropanated with the (2-furyl)carbenoid intermediate.

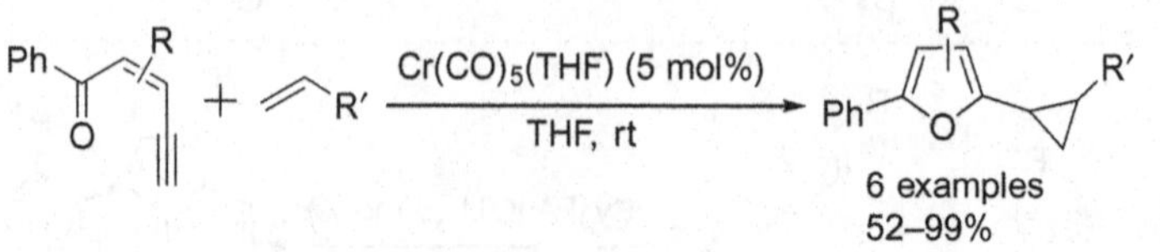

Scheme 5.85 Cr-catalyzed synthesis of 5-phenyl-2-furylcyclopropane derivatives.

In 2008, Zhang's group reported a gold-catalyzed cyclization of 1-(1-alkynyl) cyclopropyl ketones with indoles or ketones and aldehydes [148]. Through [4+2] annulation, moderate to good yields of the desired 6-membered carbo-/heterocycles can be isolated (Scheme 5.86). Additionally, imines and silyl enol ethers can be applied reaction partners as well.

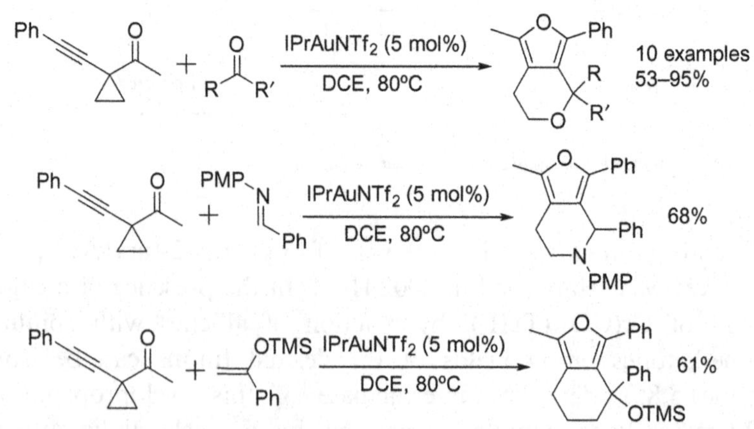

Scheme 5.86 Au-catalyzed synthesis of furans from 1-(1-alkynyl) cyclopropyl ketones and indoles.

Arcadi, Cacchi, and their co-workers studied the application of palladium catalyst in the reaction of 2-propynyl-1,3-dicarbonyls with organic halides and triflates [149]. Highly substituted furans can be produced effectively by this transformation (Scheme 5.87a). A wide variety of important functional groups on both the alkyne and the organic halide or triflate can be tolerated and better yield can be obtained by using an excess of the alkyne. From their mechanistic studies, they proved that the reaction started with intramolecular cyclization to give the vinyl palladium complex. No palladium-catalyzed Sonogashira coupling was involved. By performing the reaction under an atmosphere of carbon monoxide, the reaction affords furan derivatives incorporating carbon monoxide (Scheme 5.87b) [150]. Depending on the alkyne to organic halide or triflate ratio, acyl furans (incorporating one molecule of carbon monoxide) or enol esters (incorporating two molecules of carbon monoxide) can be isolated as the main products. This transformation was reinvestigated by Yu and Li in 2009 ($PdCl_2(PPh_3)_2$ (6 mol%), CO (13.8 bar), NEt_3, THF, 30°C) [151].

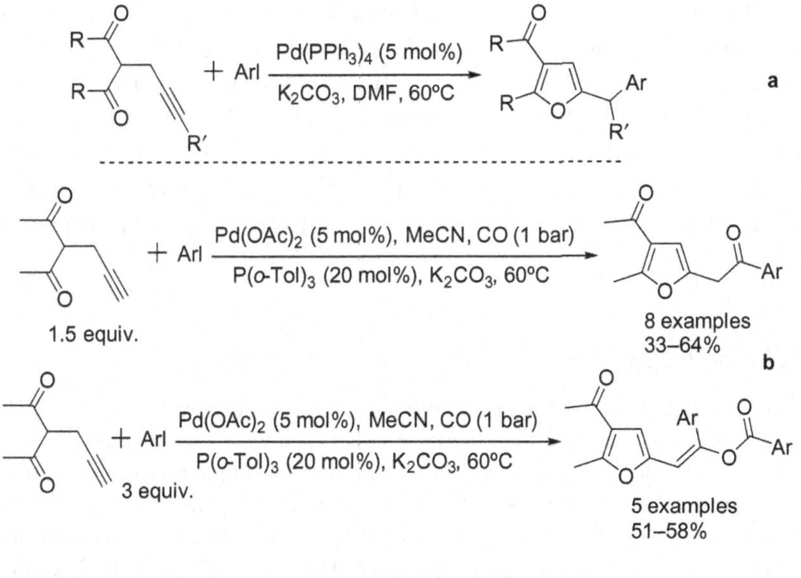

Scheme 5.87 Pd-catalyzed transformations of 2-propynyl-1,3-dicarbonyls.

In 2003, a rhodium-catalyzed procedure for the synthesis of furan-containing sulfides was reported [152]. In this procedure, the (2-furyl)carbenoids were generated from conjugated ene-yne-carbonyl compounds and then underwent carbene transfer reactions with allylic sulfides, followed by [2,3]sigmatropic rearrangement of sulfur ylides to give the final products in good yields (Scheme 5.88). When diallyl sulfide was employed, heteroatom-containing polycyclic compounds can be obtained by sequential intramolecular Diels—Alder cyclization reaction with a constructed furan ring as an enophile.

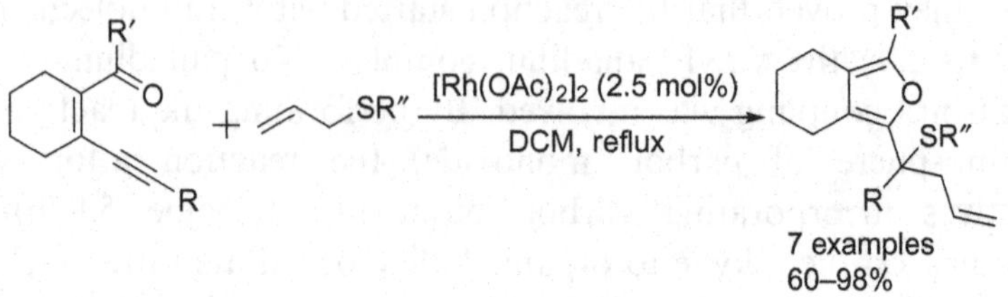

Scheme 5.88 Rh-catalyzed synthesis of furan-containing sulfides.

Recently, a Cu(I)-catalyzed cross-coupling of conjugated ene-yne-ketones and terminal alkynes to give furan-substituted allenes was reported [153]. A wide range of functional groups can be tolerated, and the products were obtained in good to excellent yields under mild conditions (Scheme 5.89a). Concerning the reaction pathway, a copper carbene migratory insertion was proposed with conjugated ene-yne-ketones as carbene precursors. Interestingly, substituted 2-furyl cyclopropene derivatives were formed by using zinc as the catalyst (Scheme 5.89b) [154]. Zinc carbenoid-like species were proposed as the intermediates which were incapable of promoting cyclopropenation reactions.

Scheme 5.89a

CuI (4.0 mg, 10 mol%) was added to a 25 ml Schlenk reaction flask. The reaction flask was then degassed for two times with nitrogen and MeCN (4 ml) was added using a syringe. Next, the conjugated ene-yne-ketones (0.22 mmol), phenylacetylene (20.4 mg, 0.20 mmol), and ethyldiisopropy-lamine (5.2 mg, 20 mol%) were successively added to the reaction solution via syringe. The resulting solution was stirred at 45°C for 10 h. The mixture was then cooled to room temperature and filtered through a short plug of silica gel (washed with petroleum ether:EtOAc = 3:1). Solvent was then removed *in vacuo* to provide a crude mixture, which was purified by silica gel column chromatography to afford pure product.

Scheme 5.89b

To a solution of the enyne (0.25 mmol) and phenylacetylene (153 mg, 1.5 mmol, 6.0 equiv.) in CH_2Cl_2 (2.5 ml, 0.1 M) at 0°C, $ZnCl_2$ (3 mg, 0.025 mmol, 10 mol%) was added. The mixture was stirred at this temperature until disappearance of enyne (checked by TLC; 6 h). The reaction mixture was then filtered through a short pad of Celite® and dried under vacuum. The resulting residue was purified by flash chromatography (deactivated SiO_2 hexanes:EtOAc = 40:1, R_f = 0.43; hexanes:EtOAc = 5:1) to afford the pure product.

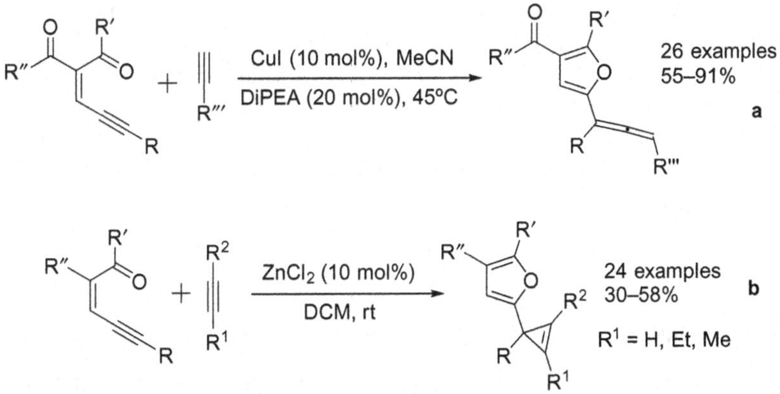

Scheme 5.89 Synthesis of furans with ZnCl₂ or CuI catalyst.

A stereocontrolled methodology for the synthesis of 2-(2-methylene-cycloalkyl)-furans was reported in 2007 [155]. By palladium-catalyzed cycloreduction of conjugated enynals bearing an alkyne unit, the corresponding furan derivatives can be produced in good to excellent yields (Scheme 5.90). Formic acid was applied as the proton source here.

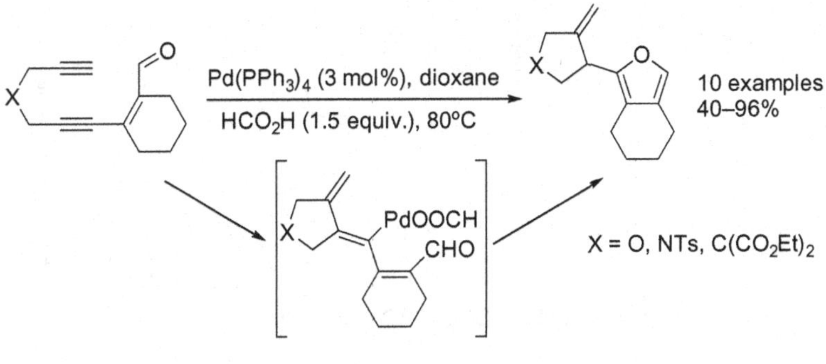

Scheme 5.90 Synthesis of 2-(2-methylenecycloalkyl)-furans.

In 2009, Zhang and co-workers reported a Rh-catalyzed carbonylative procedure for the synthesis of 5,6-dihydrocyclopenta[c]furan-4-ones [156]. The reaction proceeded in a regio- and stereo-specific manner from the corresponding 1-(1-alkynyl)cyclopropyl ketones (Scheme 5.91a). Later on, they succeeded to extend their substrates to 1-(1-alkynyl)oxiranyl ketones and gave highly substituted furo[3,4-b]furan-3(2H)-ones as the terminal products (Scheme 5.91b) [157]. The reaction proceeded via tandem heterocyclization and formal [4 + 1] cycloaddition reactions.

Scheme 5.91a

The solution of 1-(2-phenyl-1-(2-phenylethynyl)cyclopropyl)ethanone (65.0 mg, 0.25 mmol) and [{RhCl(CO)$_2$}$_2$] (4.9 mg, 0.0125 mmol) in 1,2-dichloroetane (5 ml) was degassed and then recharged with CO. Then the reaction mixture was stirred at 70°C and the TLC analysis showed the reaction was complete after 22 h. The reaction mixture was concentrated under reduced temperature and the residue was purified by column chromatography on silica gel to give the pure product.

Scheme 5.91b

The [Rh(COD)Cl]$_2$ (9.9 mg, 0.02 mmol) and S-Phos (8.2 mg, 0.02 mmol) were allowed to stir together over 1 h in 2 ml of anhydrous DCE in the presence of 1 atm of CO (balloon). The catalyst solution was then transferred to the solution of substrate (0.4 mmol) in 2 ml of anhydrous DCE in the presence of 1 atm of CO (balloon). The combined solution was then stirred at 70°C until the reaction was completed (monitored by TLC). After evaporation, the residue was purified by column chromatography on silica gel (hexanes/EtOAc = 20:1) which afforded the desired product.

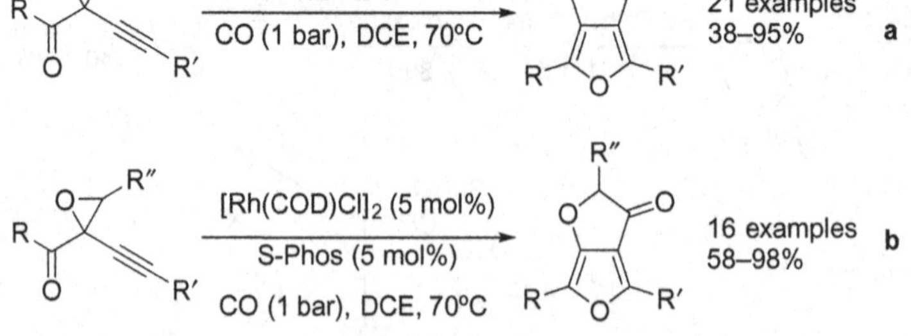

Scheme 5.91 Rh-catalyzed carbonylative synthesis of furan derivatives.

Hayashi and co-workers reported the transformation of D-glucal to furan diol in 1999 [158]. Under the assistant of a catalytic amount of $Sm(OTf)_3$ or $RuCl_2(PPh_3)_3$ in the presence of 1 equiv. of H_2O, good yield of the furan diol can be achieved under mild conditions (Scheme 5.92). In 2009, a procedure from Shaw's group for the transformation of sugars (3,4,6-tri-O-acetyl-D-glucal) into enantiomerically pure 2,3-disubstituted furans was developed [159]. $ZrCl_4$ (10 mol%) and ZnI_2 (5 mol%) were applied as the catalytic system.

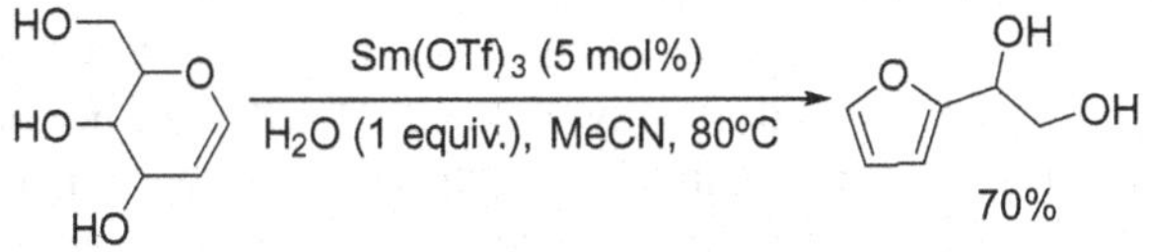

Scheme 5.92 Synthesis of furan derivative from D-glucal.

Jiang and co-workers explored the cyclization of electron-deficient alkynes and alkynols recently [160]. Here the first step is the addition of alkynols to alkynes, and then to give the final products after transition metal-catalyzed ring close reaction. Not only AgOAc/PPh$_3$ system, but also PdCl$_2$/CuI, nano-Cu$_2$O, and Fe(ClO$_4$)$_3 \cdot$ xH$_2$O catalyst systems are all effective for this transformation. Some other catalyst systems were developed as well, such as the rhodium-catalyzed system [161], AgBF$_4$-catalyzed system [162], and gold-catalyzed systems [163,164]. In 2012, Yin and co-workers reported a facile and atom-economic method for the synthesis of 3a,6a-dihydro-furo[2,3-b]furan derivatives and polysubstituted furans from furylcarbionls [165]. The reaction involved a domino Claisen rearrangement/dearomatizing electrocyclic ring-closure/aromatizing electrocyclic ring-opening sequence and gave the final products in moderate to good yields (Scheme 5.93).

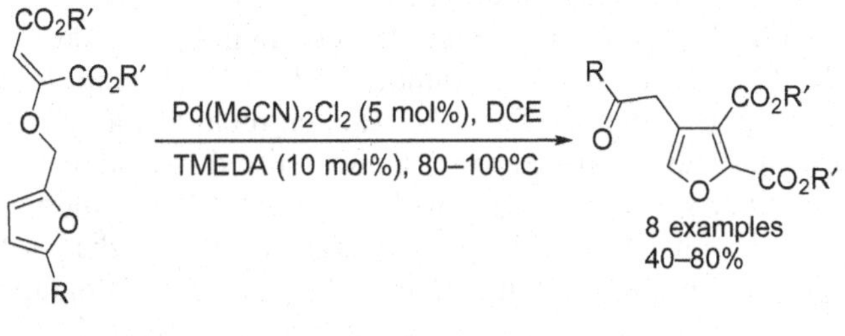

Scheme 5.93 Pd-catalyzed synthesis of furans from furylcarbionls.

A chromium-catalyzed furans production from 1,1,1-trichloroethyl propargyl ethers was established in 2002 [166]. A variety of 3-substituted furans, including such natural products as perillene and dendrolasin, were obtained in good yield via reductive annulation using $CrCl_2$ as the catalyst (Scheme 5.94). In this procedure, Mn and TMSCl were required to regenerate the Cr(II) catalyst.

Scheme 5.94

1,1,1-Trichloroethyl propargyl ether (1 mmol) in THF (2 ml) is added to a stirring, room-temperature suspension of anhydrous $CrCl_2$ (15 mol%), Mn powder (4 mmol), and freshly distilled TMSCl (4 mmol) in THF (8 ml) under argon. After complete addition, the reaction mixture was heated at 60°C. After 12–15 h, the reaction mixture was cooled to ambient, quenched with an equal volume of water, and extracted thrice with ether. The combined ethereal extracts were evaporated in vacuo and the residue was purified by SiO_2 chromatography to give 3-substituted furans in the indicated yields.

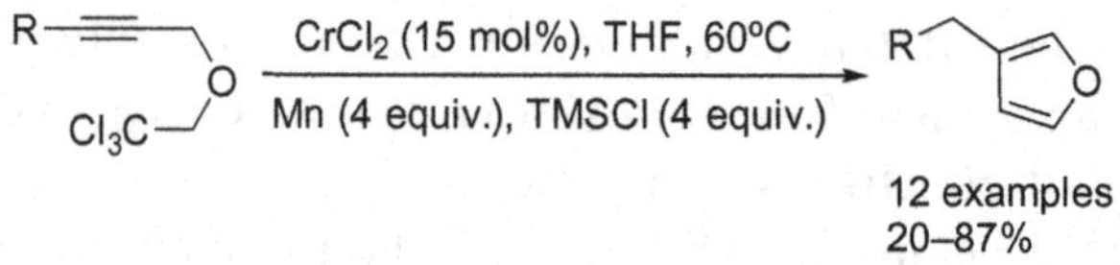

Scheme 5.94 Cr-catalyzed synthesis of furans from 1,1,1-trichloroethyl propargyl ethers.

Hammond and Xu reported a convenient procedure for the synthesis of 2,5-disubstituted furans in 2006 [167]. Started from fluoropropargyl chloride and aldehydes, good yields of the desired furans can be obtained (Scheme 5.95).

Scheme 5.95

A 50 ml round-bottom flask equipped with a magnetic stirring bar was charged with fluoropropargyl chloride (1.0 mmol), Zn dust (130 mg, 20 mmol), i-PrCHO (144 mg, 2.0 mmol), and DMF (1.0 ml), followed by $InCl_3$ (11.0 mg, 0.05 mmol). Then the reaction mixture was stirred for 1 h at room temperature, warmed to 60°C, and stirred for another 24 h. The reaction mixture was then treated with saturated NH_4Cl solution (10 ml) and extracted with ether, and the organic layer was washed with water and dried over Na_2SO_4. Purification with flash chromatography (pure hexane to 10% ethyl acetate in hexane) gave pure compound.

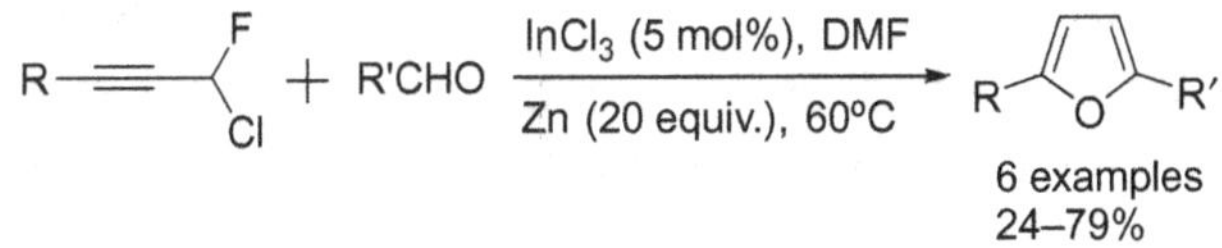

Scheme 5.95 In-catalyzed synthesis of furans from fluoropropargyl chloride.

Li and co-workers reported the synthesis of furans from aryloxye-nynes in 2012 [168]. The substrates, (*E*)-1-aryloxy-1-en-3-ynes, were prepared by Sonogashira coupling of 2-bromo-3-aryloxypropenoates with terminal alkynes using Pd(PPh$_3$)$_4$ and CuI as the catalysts in Et$_3$N. Under the assistant of gold catalyst, enynyl-aryl ethers were transformed into 2,4-disubstituted furans with an ester group at C-4 position (Scheme 5.96a). Meanwhile, they also found that 2,3,4-trisubstituted furans can be produced effectively by palladium-catalyzed reaction from aryloxy-enynes and aryl iodides in good to excellent yields (Scheme 5.96b) [169].

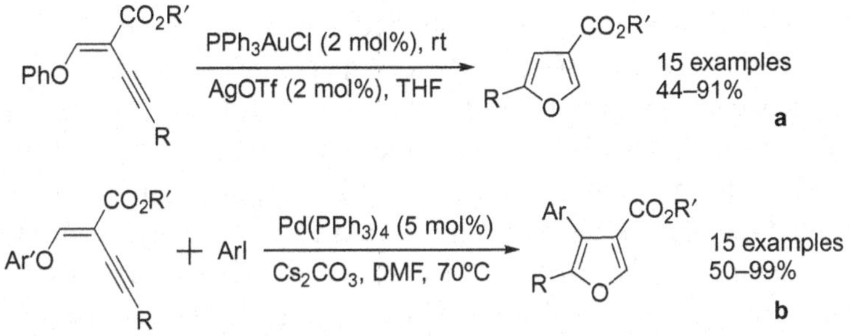

Scheme 5.96 Synthesis of furans from enynyl-aryl ethers.

Recently, Reddy and co-workers developed the synthesis of 5-substituted furan-3-carboxylates from Morita−Baylis−Hillman acetates of acetylenic aldehydes [170]. This process involves palladium-catalyzed isomerization followed by base-promoted deacetylative 5-exo-dig-cycloisomerization reactions. Various 3-furanoates were produced in good yields and furanones can be produced as well (Scheme 5.97a). The use of 1,3-dienyl alkyl ethers as substrates and *P*-toluenesulfonic acid as the catalyst for furan synthesis was developed as well [171]. Later on, it was found that enals can also be applied as substrates for furan synthesis under palladium-catalyzed oxidative conditions [172]. Additionally, the use of carbonyl-ene-nitrile compounds and propargyl carboxylates as starting materials for furans synthesis was developed in 2010 [173]. In the presence of a catalytic amount of $PtCl_2$, α-alkylidene-N-furylimines can be produced in good yields with high stereoselectivities (Scheme 5.97b).

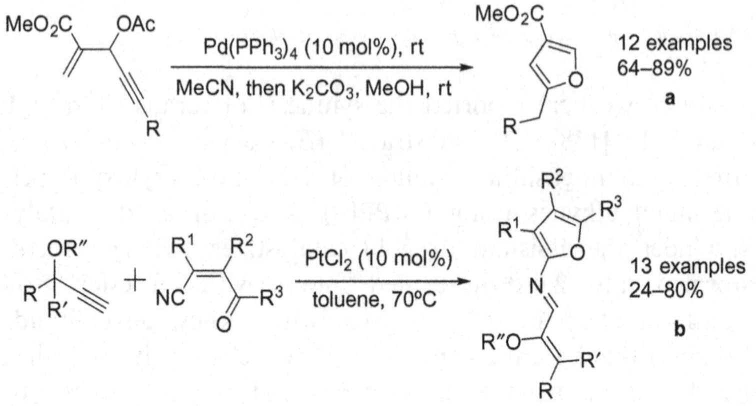

Scheme 5.97 Synthesis of furans from alkynes.

In 1990, Müller and Pautex studied the behavior of rhodium (II) catalyst in the isomerization of cyclopropenes. Furans can be obtained via vinylcarbene intermediates [174]. Padwa and co-workers reported their achievements in 1991 on this rhodium (I)-catalyzed transformation [175]. They found that Rh(II)-catalyzed reaction of unsymmetrically substituted cyclopropenes gave substituted furans derived from cleavage of the less substituted δ-bond. In contrast, Rh(I) catalyst resulted the cleavage of the more substituted δ-bond. In 2003, a system based on cheaper palladium catalyst was reported by Ma and Zhang (Scheme 5.98) [176]. 2,3,4-Trisubstituted furans or 2,3,5-trisubstituted furans can be produced regioselectively by cycloisomerization of cyclopropenyl ketones in the presence of CuI or $PdCl_2(CH_3CN)_2$, respectively.

Scheme 5.98

$PdCl_2(CH_3CN)_2$ (6.5 mg, 5 mol%) was added to a solution of substrate (0.5 mmol) in 2 ml of $CHCl_3$. The mixture was then stirred under reflux for 18 h. Evaporation and chromatography on silica gel (petroleum ether/ diethyl ether 10:1) gave the pure product.

Or CuI (4.7 mg, 5 mol%) was added to a solution of substrate (0.5 mmol) in 2 ml of CH_3CN. The mixture was then stirred at 80°C for 9 h. Evaporation and chromatography on silica gel (petroleum ether/ diethyl ether 10:1) afforded the pure product.

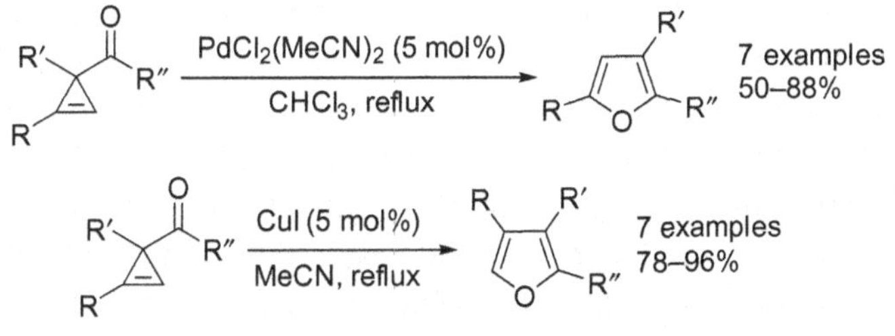

Scheme 5.98 Synthesis of furans from cyclopropenyl ketones.

Xu and co-workers found that cyclopropenes can be dimerized to bifurans by using palladium and copper as the combinational catalyst (Scheme 5.99) [177]. The reaction proceeded via copper-promoted cycloisomerization and palladium-catalyzed dimerization cascade reactions. These novel bifuran structures possess interesting optoelectronic properties. Furan palladium complex was found to be the intermediate. Hence, they also developed two other procedures for the synthesis of furans. Tetrasubstituted furan carboxylates can be prepared using a tandem cycloisomerization/oxidative carbonylation sequence of cyclopropenes [178], and tetrasubstituted alkene-functionalized furans can be prepared by isomerization/olefination cascade sequence using copper/palladium relay catalysis [179].

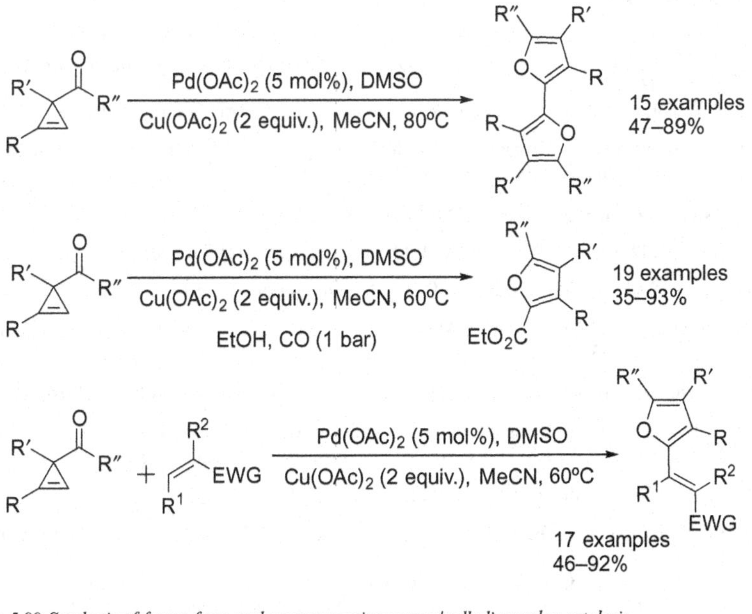

Scheme 5.99 Synthesis of furans from cyclopropenes using copper/palladium relay catalysis.

Maas and co-workers reported a copper(I) chloride-catalyzed trans-formation of α-[(2-alkynyl)oxy]silyl-α-diazoacetates (containing terminal alkyne functions) to the corresponding bicyclic 2-methoxyfurans in 1999 (Scheme 5.100a) [180]. In 2009, Barluenga and co-workers reported a copper(I)-catalyzed synthesis of furyl-substituted cyclobutenes from vinyldiazoacetates and propargylic esters (Scheme 5.100b) [181]. Moreover, ruthenium-catalyzed olefin-metathesis was applied in furan derivatives synthesis as well [182].

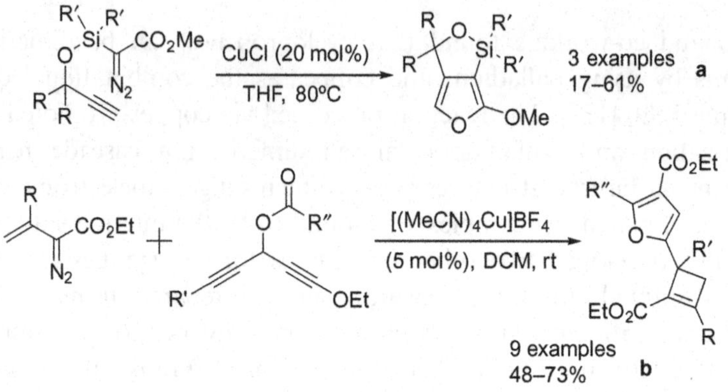

Scheme 5.100 Cu-catalyzed synthesis of furans from diazo compounds.

REFERENCES

[1] a. B. Seiller, C. Bruneau, P.H. Dixneuf, Tetrahedron 51 (1995) 13089–13102.
    b. H. Kücükbay, B. Cetinkaya, S. Guesmi, P.H. Dixneuf, Organometallics 15 (1996) 2434–2439.
    c. B. Seiller, C. Bruneau, P.H. Dixneuf, J. Chem. Soc. Chem. Commun. (1994) 493–494.

[2] A.E. Díaz-Álvarez, P. Crochet, M. Zablocka, C. Duhayon, V. Cadierno, J. Gimeno, et al., Adv. Synth. Catal. 348 (2006) 1671–1679.

[3] D. Végh, P. Zalupsky, J. Kovác, Synth. Commun. 20 (1990) 1113–1123.

[4] S. Elgafi, L.D. Field, B.A. Messerle, J. Organomet. Chem. 607 (2000) 97–104.

[5] B. Gabriele, G. Salerno, E. Lauria, J. Org. Chem. 64 (1999) 7687–7692.

[6] V. Cadierno, J. Díez, J. García-Álvarez, J. Gimeno, N. Nebra, J. Rubio-García, Dalton Trans. (2006) 5593–5604.

[7] C.C. Schneider, H. Caldeira, B.M. Gay, D.F. Back, G. Zeni, Org. Lett. 12 (2010) 936–939.

[8] a. Y. Liu, F. Song, Z. Song, M. Liu, B. Yan, Org. Lett. 7 (2005) 5409–5412.
    b. X. Du, F. Song, Y. Lu, H. Chen, Y. Liu, Tetrahedron 65 (2009) 1839–1845.

[9] C. Praveen, P. Kiruthiga, P.T. Perumal, Synlett (2009) 1990–1996.

[10] a. F.E. McDonald, C.B. Connolly, M.M. Gleason, T.B. Towne, K.D. Treiber, J. Org. Chem. 58 (1993) 6952–6953.
     b. F.E. McDonald, C.C. Schultz, J. Am. Chem. Soc. 116 (1994) 9363–9364.
     c. F.E. McDonald, M.M. Gleason, J. Am. Chem. Soc. 118 (1996) 6648–6659.

[11] J.A. Marshall, C.A. Sehon, J. Org. Chem. 60 (1995) 5966–5968.

[12] F.L. Qing, W.Z. Gao, J. Ying, J. Org. Chem. 65 (2000) 2003–2006.

[13] D. Zhang, C. Yuan, Eur. J. Org. Chem. (2007) 3916–3924.

[14] S. Kim, P.H. Lee, Adv. Synth. Catal. 350 (2008) 547–551.

[15] A.S.K. Hashmi, T. Häffner, M. Rudolph, F. Rominger, Eur. J. Org. Chem. (2011) 667–671.

[16] J. Zhang, X. Zhao, L. Lu, Tetrahedron Lett. 48 (2007) 1911–1913.

[17] a. B. Gabriele, G. Salerno, F. De Pascali, G.T. Scianò, M. Costa, G.P. Chiusoli, Tetrahedron Lett. 38 (1997) 6877–6880.
      b. B. Gabriele, G. Salerno, F. De Pascali, M. Costa, G.P. Chiusoli, J. Org. Chem. 64 (1999) 7693–7699.

[18] B. Gabriele, P. Plastina, G. Salerno, R. Mancuso, Synthesis (2006) 4247–4251.

[19] B. Gabriele, L. Veltri, R. Mancuso, P. Plastina, G. Salerno, M. Costa, Tetrahedron Lett. 51 (2010) 1663–1665.

[20] F. Camus, B. Hasiak, M. Martin, D. Couturier, Synth. Commun. 12 (1982) 647–650.

[21] J. Ji, X. Lu, J. Chem. Soc. Chem. Commun. (1993) 764–765.

[22] a. S.J. Pridmore, P.A. Slatford, J.M.J. Williams, Tetrahedron Lett. 48 (2007) 5111–5114.
      b. S.J. Pridmore, P.A. Slatford, J.E. Taylor, M.K. Whittlesey, J.M.J. Williams, Tetrahedron 65 (2009) 8981–8986.

[23] K. Tanaka, T. Shoji, M. Hirano, Eur. J. Org. Chem. (2007) 2687–2699.

[24] S.J. Hayes, D.W. Knight, M.D. Menzies, M. O'Halloran, W.F. Tan, Tetrahedron Lett. 48 (2007) 7709–7712.

[25] S.J. Hayes, D.W. Knight, A.W.T. Smith, M.J. O'Halloran, Tetrahedron Lett. 51 (2010) 717–719.

[26] S. Kim, D. Kang, S. Shin, P.H. Lee, Tetrahedron Lett. 51 (2010) 1899–1901.

[27] B. Gabriele, P. Plastina, M.V. Vetere, L. Veltri, R. Mancuso, G. Salerno, Tetrahedron Lett. 51 (2010) 3565–3567.

[28] X. Zhang, Z. Lu, C. Fu, S. Ma, J. Org. Chem. 75 (2010) 2589–2598.

[29] J.M. Aurrecoechea, A. Durana, E. Pérez, J. Org. Chem. 73 (2008) 3650–3653.

[30] M. Doe, T. Shibue, H. Haraguchi, Y. Morimoto, Org. Lett. 7 (2005) 1765–1768.

[31] Y. Yada, Y. Miyake, Y. Nishibayashi, Organometallics 27 (2008) 3614–3617.

[32] M. Egi, K. Azechi, S. Akai, Org. Lett. 11 (2009) 5002–5005.

[33] L. Zhu, J. Luo, R. Hong, Org. Lett. 16 (2014) 2162–2165.

[34] A. Aponick, C.Y. Li, J. Malinge, E.F. Marques, Org. Lett. 11 (2009) 4624–4627.

[35] P.A. Allegretti, E.M. Ferreira, Org. Lett. 13 (2011) 5924–5927.

[36] K. Ravindar, M.S. Reddy, P. Deslongchamps, Org. Lett. 13 (2011) 3178–3181.

[37] H.H. Tso, H. Tsay, Tetrahedron Lett. 38 (1997) 6869–6870.

[38] Y. Wakabayashi, Y. Fukuda, H. Shiragami, K. Utimoto, H. Nozaki, Tetrahedron Lett. 41 (1985) 3655–3661.

[39] M. Sakai, M. Sasaki, K. Tanino, M. Miyashita, Tetrahedron Lett. 43 (2002) 1705–1708.

[40] S. Wen, W. Liu, Y. Liang, Synthesis (2007) 3295–3300.

[41] M.N. Pennell, R.W. Foster, P.G. Turner, H.C. Hailes, C.J. Tame, T.D. Sheppard, Chem. Commun. 50 (2014) 1302–1304.

[42] D. Miller, J. Chem. Soc. (C) (1969) 12–15.

[43] C.M. Marson, S. Harper, R. Wrigglesworth, J. Chem. Soc., Chem. Commun. (1994) 1879–1880.

[44] a. J.M. Aurrecoechea, E. Pérez, Tetrahedron Lett. 42 (2001) 3839–3841.
     b. J.M. Aurrecoechea, E. Pérez, M. Solay, J. Org. Chem. 66 (2001) 564–569.
     c. J.M. Aurrecoechea, E. Pérez, Tetrahedron 60 (2004) 4139–4149.

[45] A.S.K. Hashmi, P. Sinha, Adv. Synth. Catal. 346 (2004) 432–438.

[46] X.-Z. Shu, X.-Y. Liu, H.-Q. Xiao, K.-G. Ji, L.-N. Guo, C.-Z. Qi, et al., Adv. Synth. Catal. 349 (2007) 2493–2498.

[47] K.-G. Ji, Y.-W. Shen, X.-Z. Shu, H.-Q. Xiao, Y.-J. Bian, Y.-M. Liang, Adv. Synth. Catal. 350 (2008) 1275–1280.

[48] K.-G. Ji, X.-Z. Shu, J. Chen, S.-C. Zhao, Z.-J. Zheng, X.-Y. Liu, et al., Org. Biomol. Chem. 7 (2009) 2501–2505.

[49] L.-Z. Dai, M. Shi, Tetrahedron Lett. 49 (2008) 6437–6439.

[50] R.K. Shiroodi, O. Koleda, V. Gevorgyan, J. Am. Chem. Soc. 136 (2014) 13146–13149.

[51] A. Blanc, A. Alix, J.-M. Weibel, P. Pale, Eur. J. Org. Chem. (2010) 1644–1647.

[52] A. Blanc, K. Tenbrink, J.-M. Weibel, P. Pale, J. Org. Chem. 74 (2009) 5342–5348.

[53] A. Blanc, K. Tenbrink, J.-M. Weibel, P. Pale, J. Org. Chem. 74 (2009) 4360–4363.

[54] a. M. Yoshida, M. Al-Amin, K. Matsuda, K. Shishido, Tetrahedron Lett. 49 (2008) 5021–5023.
     b. M. Yoshida, M. Al-Amin, K. Shishido, Synthesis (2009) 2454–2466.

[55] C.-Y. Lo, H. Guo, J.-J. Lian, F.-M. Shen, R.-S. Liu, J. Org. Chem. 67 (2002) 3930–3932.

[56] J.Y. Kang, B.T. Connell, J. Org. Chem. 76 (2011) 2379–2383.

[57] C.-Y. Zhou, P.W.H. Chan, C.-M. Che, Org. Lett. 8 (2006) 325–328.

[58] A.W. Sromek, M. Rubina, V. Gevorgyan, J. Am. Chem. Soc. 127 (2005) 10500–10501.

[59] Y. Xia, A.S. Dudnik, V. Gevorgyan, Y. Li, J. Am. Chem. Soc. 130 (2008) 6940–6941.

[60] A.S. Dudnik, Y. Xia, Y. Li, V. Gevorgyan, J. Am. Chem. Soc. 132 (2010) 7645–7655.

[61] A.S. Dudnik, V. Gevorgyan, Angew. Chem. Int. Ed. 46 (2007) 5195–5197.

[62] J.A. Marshall, X.J. Wang, J. Org. Chem. 56 (1991) 960–969.

[63] J.A. Marshall, G.S. Bartley, J. Org. Chem. 59 (1994) 7169–7171.

[64] a. J.A. Marshall, E.M. Wallace, J. Org. Chem. 60 (1995) 796–797.
     b. J.A. Marshall, X.J. Wang, J. Org. Chem. 56 (1991) 6264–6266.
     c. J.A. Marshall, X.J. Wang, J. Org. Chem. 57 (1992) 3387–3396.
     d. J.A. Marshall, E.D. Robinson, J. Org. Chem. 55 (1990) 3450–3451.
     e. C.M. Marson, S. Harper, J. Org. Chem. 63 (1998) 9223–9231.
     f. J.A. Marshall, G.S. Bartley, E.M. Wallace, J. Org. Chem. 61 (1996) 5729–5735.

[65] X. Cong, K.-G. Liu, Q.-J. Liao, Z.-J. Yao, Tetrahedron Lett. 46 (2005) 8567–8571.

[66] A.S.K. Hashmi, J.H. Choi, J.W. Bats, J. Prakt. Chem. 341 (1999) 341–357.

[67] A.S.K. Hashmi, L. Schwarz, J.W. Bats, J. Prakt. Chem. 342 (2000) 40–51.

[68] X. Yu, J. Zhang, Adv. Synth. Catal. 353 (2011) 1265–1268.

[69] M. Miao, X. Xu, L. Xu, H. Ren, Eur. J. Org. Chem. (2014) 5896–5900.

[70] J.T. Kim, A.V. Kel'in, V. Gevorgyan, Angew. Chem. Int. Ed. 42 (2003) 98–101.

[71] H.Y. Kim, J.-Y. Li, K. Oh, J. Org. Chem. 77 (2012) 11132–11145.

[72] A.W. Sromek, A.V. Kel'in, V. Gevorgyan, Angew. Chem. Int. Ed. 43 (2004) 2280–2282.

[73] T. Schwier, A.W. Sromek, D.M.L. Yap, D. Chernyak, V. Gevorgyan, J. Am. Chem. Soc. 129 (2007) 9868–9878.

[74] L. Peng, X. Zhang, M. Ma, J. Wang, Angew. Chem. Int. Ed. 46 (2007) 1905–1908.

[75] E. Wang, X. Fu, X. Xie, J. Chen, H. Gao, Y. Liu, Tetrahedron Lett. 52 (2011) 1968–1972.

[76] Y. Deng, C. Fu, S. Ma, Chem. Eur. J. 17 (2011) 4976–4980.

[77] C. Cheng, S. Liu, G. Zhu, Org. Lett. 17 (2015) 1581–1584.

[78] A.S.K. Hashmi, Angew. Chem. Int. Ed. 34 (1995) 1581–1583.

[79] A.S.K. Hashmi, T.L. Ruppert, T. Knöfel, J.W. Bats, J. Org. Chem. 62 (1997) 7295–7304.

[80] A.S.K. Hashmi, L. Schwarz, M. Bolte, Eur. J. Org. Chem. (2004) 1923–1935.

[81] B. Alcaide, P. Almendros, T.M. del Campo, Eur. J. Org. Chem. (2007) 2844–2849.

[82] S. Ma, Z. Yu, Angew. Chem. Int. Ed. 41 (2002) 1775–1778.

[83] S. Ma, Z. Yu, Chem. Eur. J. 10 (2004) 2078–2087.

[84] S. Ma, Z. Gu, Z. Yu, J. Org. Chem. 70 (2005) 6291–6294.

[85] S. Ma, J. Zhang, Chem. Commun. (2000) 117–118.

[86] S. Ma, J. Zhang, L. Lu, Chem. Eur. J. 9 (2003) 2447–2456.

[87] S. Ma, L. Li, Org. Lett. 2 (2000) 941–944.

[88] K. Kato, T. Mochida, H. Takayama, M. Kimura, H. Moriyama, A. Takeshita, et al., Tetrahedron Lett. 50 (2009) 4744–4746.

[89] J. Reisch, A. Bathe, Liebigs Ann. Chem. (1988) 69–73.

[90] A.V. Kel'in, V. Gevorgyan, J. Org. Chem. 67 (2002) 95–98.

[91] a. H. Sheng, S. Lin, Y. Huang, Tetrahedron Lett. 27 (1986) 4893–4894.
      b. H. Sheng, S. Lin, Y. Huang, Synthesis (1987) 1022–1023.

[92] Y. Fukuda, H. Shiragami, K. Utimoto, H. Nozaki, J. Org. Chem. 56 (1991) 5816–5819.

[93] a. A. Jeevanandam, K. Narkunan, Y.C. Ling, J. Org. Chem. 66 (2001) 6014–6020.
      b. A. Jeevanandam, K. Narkunan, C. Cartwright, Y.C. Ling, Tetrahedron Lett. 40 (1999) 4841–4844.

[94] a. A. Sniady, A. Durham, M.S. Morreale, K.A. Wheeler, R. Dembinski, Org. Lett. 9 (2007) 1175–1178.
      b. A. Sniady, A. Durham, M.S. Morreale, A. Marcinek, S. Szafert, T. Lis, et al., J. Org. Chem. 73 (2008) 5881–5889.

[95] a. H. Imagawa, T. Kurisaki, M. Nishizawa, Org. Lett. 6 (2004) 3679–3681.
      b. D. Ménard, A. Vidal, C. Barthomeuf, J. Lebreton, P. Gosselin, Synlett (2006) 57–60.

[96] V. Aucagne, F. Amblard, L.A. Agrofoglio, Synlett (2004) 2406–2408.

[97] a. R.H.E. Hudson, J.M. Moszynski, Synlett (2006) 2997–3000.
      b. N. Esho, J.-P. Desaulniers, B. Davies, H.M.-P. Chui, M.S. Rao, C.S. Chow, et al., Bioorg. Med. Chem. 13 (2005) 1231–1238.
      c. F. Amblard, V. Aucagne, P. Guenot, R.F. Schinazi, L.A. Agrofoglio, Bioorg. Med. Chem. 13 (2005) 1239–1248.

[98] P. Wipf, L.T. Rahman, S.R. Rector, J. Org. Chem. 63 (1998) 7132–7133.

[99] a. P. Wipf, M.J. Soth, Org. Lett. 4 (2002) 1787–1790.
      b. P. Wipf, M. Grenon, Can. J. Chem. 84 (2006) 1226–1241.

[100] V. Belting, N. Krause, Org. Biomol. Chem. 7 (2009) 1221−1225.

[101] T. Wang, J. Zhang, Dalton Trans. 39 (2010) 4270−4273.

[102] Y. Liu, Z. Liu, Y. Cui, Chin. J. Chem. 33 (2015) 175−180.

[103] W. Li, J. Zhang, Chem. Commun. 46 (2010) 8839−8841.

[104] M.E. Krafft, D.V. Vidhani, J.W. Cran, M. Manoharan, Chem. Commun. 47 (2011) 6707−6709.

[105] K.-D. Umland, A. Palisse, T.T. Haug, S.F. Kirsch, Angew. Chem. Int. Ed. 50 (2011) 9965−9968.

[106] H.S.P. Rao, S. Jothilingam, J. Org. Chem. 68 (2003) 5392−5394.

[107] X. Han, R.A. Widenhoefer, J. Org. Chem. 69 (2004) 1738−1740.

[108] a. Y. Li, K.A. Wheeler, R. Dembinski, Eur. J. Org. Chem. (2011) 2767−2771.
       b. Y. Li, K.A. Wheeler, R. Dembinski, Org. Biomol. Chem. 10 (2012) 2395−2408.

[109] W.B. Zhang, S.D. Xiu, C.Y. Li, Org. Chem. Front. 2 (2015) 47−50.

[110] G. Zhou, F. Liu, J. Zhang, Chem. Eur. J. 17 (2011) 3101−3104.

[111] K. Mal, A. Sharma, I. Das, Chem. Eur. J. 20 (2014) 11932−11945.

[112] C.H. Oh, V.R. Reddy, A. Kim, C.Y. Rhim, Tetrahedron Lett. 47 (2006) 5307−5310.

[113] X. Liu, Z. Pan, X. Shu, X. Duan, Y. Liang, Synlett (2006) 1962−1964.

[114] R. Jana, S. Paul, A. Biswas, J.K. Ray, Tetrahedron Lett. 51 (2010) 273−276.

[115] L. Chen, Y. Fang, Q. Zhao, M. Shi, C. Li, Tetrahedron Lett. 51 (2010) 3678−3681.

[116] A. Rodríguez, W.J. Moran, Tetrahedron Lett. 52 (2011) 2605−2607.

[117] R.K. Shiroodi, C.I.R. Vera, A.S. Dudnik, V. Gevorgyan, Tetrahedron Lett. 56 (2015) 3251−3254.

[118] N.D. Shapiro, F.D. Toste, J. Am. Chem. Soc. 129 (2007) 4160−4161.

[119] a. J. Barluenga, L. Riesgo, R. Vicente, L.A. López, M. Tomás, J. Am. Chem. Soc. 130 (2008) 13528−13529.
       b. J. Barluenga, L. Riesgo, R. Vicente, L.A. López, M. Tomás, J. Am. Chem. Soc. 129 (2007) 7772−7773.

[120] S. Ma, L. Lu, J. Zhang, J. Am. Chem. Soc. 126 (2004) 9645−9660.

[121] a. A. Padwa, F.R. Kinder, J. Org. Chem. 58 (1993) 21−28.
       b. A. Padwa, D.C. Dean, D.J. Fairfax, S.L. Xu, J. Org. Chem. 58 (1993) 4646−4655.
       c. A. Padwa, U. Chiacchio, Y. Garreau, J.M. Kassir, K.E. Krumpe, A.M. Schoffstall, J. Org. Chem. 55 (1990) 414−416.
       d. A. Padwa, J. Organometal. Chem. 610 (2000) 88−101.
       e. F.R. Kinder, A. Padwa, Tetrahedron Lett. 31 (1990) 6835−6838.

[122] T.R. Hoye, C.J. Dinsmore, D.S. Johnson, P.F. Korkowski, J. Org. Chem. 55 (1990) 4518−4520.

[123] C.H. Oh, S.J. Lee, J.H. Lee, Y.J. Na, Chem. Commun. (2008) 5794−5796.

[124] G. Zhou, J. Zhang, Chem. Commun. 46 (2010) 6593−6595.

[125] M.Y. Chang, Y.C. Cheng, Y.J. Lu, Org. Lett. 17 (2015) 1264−1267.

[126] K. Miki, Y. Washitake, K. Ohe, S. Uemura, Angew. Chem. Int. Ed. 43 (2004) 1857−1860.

[127] S. Ma, J. Zhang, Angew. Chem. Int. Ed. 42 (2003) 183−187.

[128] J. Zhang, H.G. Schmalz, Angew. Chem. Int. Ed. 45 (2006) 6704−6707.

[129] Y. Xiao, J. Zhang, Angew. Chem. Int. Ed. 47 (2008) 1903−1906.

[130] Y. Xiao, J. Zhang, Adv. Synth. Catal. 351 (2009) 617–629.

[131] L. Zhao, G. Cheng, Y. Hu, Tetrahedron Lett. 49 (2008) 7364–7367.

[132] R. Liu, J. Zhang, Chem. Eur. J. 15 (2009) 9303–9306.

[133] F. Liu, Y. Yu, J. Zhang, Angew. Chem. Int. Ed. 48 (2009) 5505–5508.

[134] F. Liu, D. Qian, L. Li, X. Zhao, J. Zhang, Angew. Chem. Int. Ed. 49 (2010) 6669–6672.

[135] L. Zhou, M. Zhang, W. Li, J. Zhang, Angew. Chem. Int. Ed. 53 (2014) 6542–6545.

[136] Y. Bai, J. Fang, J. Ren, Z. Wang, Chem. Eur. J. 15 (2009) 8975–8978.

[137] T. Yao, X. Zhang, R.C. Larock, J. Am. Chem. Soc. 126 (2004) 11164–11165.

[138] T. Yao, X. Zhang, R.C. Larock, J. Org. Chem. 70 (2005) 7679–7685.

[139] B.V.S. Reddy, V.V. Reddy, G. Karthik, B. Jagadeesh, Tetrahedron 71 (2015) 2572–2578.

[140] N.T. Patil, H. Wu, Y. Yamamoto, J. Org. Chem. 70 (2005) 4531–4534.

[141] W. Zhao, J. Zhang, Chem. Commun. 46 (2010) 7816–7818.

[142] W. Zhao, J. Zhang, Chem. Commun. 46 (2010) 4384–4386.

[143] W. Zhao, J. Zhang, Org. Lett. 13 (2011) 688–691.

[144] H. Gao, X. Zhao, Y. Yu, J. Zhang, Chem. Eur. J. 16 (2010) 456–459.

[145] H. Gao, X. Wu, J. Zhang, Chem. Eur. J. 17 (2011) 2838–2841.

[146] A. Saito, Y. Enomoto, Y. Hanzawa, Tetrahedron Lett. 52 (2011) 4299–4302.

[147] a. K. Miki, F. Nishino, K. Ohe, S. Uemura, J. Am. Chem. Soc. 124 (2002) 5260–5261.
       b. K. Miki, T. Yokoi, F. Nishino, Y. Kato, Y. Washitake, K. Ohe, et al., J. Org. Chem. 69 (2004) 1557–1564.
       c. K. Miki, T. Yokoi, F. Nishino, K. Ohe, S. Uemura, J. Organometal. Chem. 645 (2002) 228–234.

[148] G. Zhang, X. Huang, G. Li, L. Zhang, J. Am. Chem. Soc. 130 (2008) 1814–1815.

[149] a. S. Cacchi, G. Fabrizi, L. Moro, J. Org. Chem. 62 (1997) 5327–5332.
       b. A. Arcadi, S. Cacchi, G. Fabrizi, F. Marinellia, L.M. Parisi, Tetrahedron 59 (2003) 4661–4671.
       c. A. Arcadi, S. Cacchi, R.C. Larock, F. Marinelli, Tetrahedron Lett. 34 (1993) 2813–2816.

[150] A. Arcadi, E. Rossi, Tetrahedron Lett. 37 (1996) 6811–6814.

[151] Y. Li, Z. Yu, J. Org. Chem. 74 (2009) 8904–8907.

[152] Y. Kato, K. Miki, F. Nishino, K. Ohe, S. Uemura, Org. Lett. 5 (2003) 2619–2621.

[153] F. Hu, Y. Xia, C. Ma, Y. Zhang, J. Wang, Org. Lett. 16 (2014) 4082–4085.

[154] M.J. González, L.A. Loópez, R. Vicente, Org. Lett. 16 (2014) 5780–5783.

[155] C.H. Oh, H.M. Park, D.I. Park, Org. Lett. 9 (2007) 1191–1193.

[156] Y. Zhang, Z. Chen, Y. Xiao, J. Zhang, Chem. Eur. J. 15 (2009) 5208–5211.

[157] T. Wang, C.-H. Wang, J. Zhang, Chem. Commun. 47 (2011) 5578–5580.

[158] M. Hayashi, H. Kawabata, K. Yamada, Chem. Commun. (1999) 965–966.

[159] M. Saquib, I. Husain, B. Kumar, A.K. Shaw, Chem. Eur. J 15 (2009) 6041–6049.

[160] a. H. Cao, H. Jiang, R. Mai, S. Zhu, C. Qi, Adv. Synth. Catal. 352 (2010) 143–152.
       b. H. Cao, H. Jiang, H. Huang, J. Zhao, Org. Biomol. Chem. 9 (2011) 7313–7317.
       c. H. Cao, H. Jiang, G. Yuan, Z. Chen, C. Qi, H. Huang, Chem. Eur. J. 16 (2010) 10553–10559.
       d. H. Jiang, W. Yao, H. Cao, H. Huang, D. Cao, J. Org. Chem. 75 (2010) 5347–5350.

[161] H. Kawai, S. Oi, Y. Inoue, Heterocycles 67 (2006) 101–105.

[162] A. Palisse, S.F. Kirsch, Eur. J. Org. Chem. (2014) 7095–7098.

[163] F.M. Istrate, F. Gagosz, Beilstein J. Org. Chem. 7 (2011) 878–885.

[164] M.H. Suhre, M. Reif, S.F. Kirsch, Org. Lett. 7 (2005) 3925–3927.

[165] B. Yin, G. Zeng, C. Cai, F. Ji, L. Huang, Z. Li, et al., Org. Lett. 14 (2012) 616–619.

[166] D.K. Barma, A. Kundu, R. Baati, C. Mioskowski, J.R. Falck, Org. Lett. 4 (2002) 1387–1389.

[167] B. Xu, G.B. Hammond, J. Org. Chem. 71 (2006) 3518–3521.

[168] E. Li, W. Yao, X. Xie, C. Wang, Y. Shao, Y. Li, Org. Biomol. Chem. 10 (2012) 2960–2965.

[169] E. Li, X. Cheng, C. Wang, Y. Shao, Y. Li, J. Org. Chem. 77 (2012) 7744–7748.

[170] C.R. Reddy, G. Krishna, M.D. Reddy, Org. Biomol. Chem. 12 (2014) 1664–1670.

[171] M. Zhang, H.-F. Jiang, H. Neumann, M. Beller, P.H. Dixneuf, Angew. Chem. Int. Ed. 48 (2009) 1681–1684.

[172] M.G. Lauer, W.H. Henderson, A. Awad, J.P. Stambuli, Org. Lett. 14 (2012) 6000–6003.

[173] M. Murai, S. Yoshida, K. Miki, K. Ohe, Chem. Commun. 46 (2010) 3366–3368.

[174] a. P. Müller, N. Pautex, M.P. Doyle, V. Bagheri, Helv. Chim. Acta 73 (1990) 1233–1241.
      b. P. Müller, C. Gränicher, Helv. Chim. Acta 78 (1995) 129–144.

[175] A. Padwa, J.M. Kassir, S.L. Xu, J. Org. Chem. 56 (1991) 6971–6972.

[176] S. Ma, J. Zhang, J. Am. Chem. Soc. 125 (2003) 12386–12387.

[177] C. Song, D. Sun, X. Peng, J. Bai, R. Zhang, S. Hou, et al., Chem. Commun. 49 (2013) 9167–9169.

[178] C. Song, S. Dong, L. Feng, X. Peng, M. Wang, J. Wang, et al., Org. Biomol. Chem. 11 (2013) 6258–6262.

[179] C. Song, L. Ju, M. Wang, P. Liu, Y. Zhang, J. Wang, et al., Chem. Eur. J. 19 (2013) 3584–3589.

[180] V. Gettwert, F. Krebs, G. Maas, Eur. J. Org. Chem. (1999) 1213–1221.

[181] J. Barluenga, L. Riesgo, L.A. López, E. Rubio, M. Tomás, Angew. Chem. Int. Ed. 48 (2009) 7569–7572.

[182] a. T.J. Donohoe, A.J. Orr, K. Gosby, M. Bingham, Eur. J. Org. Chem. (2005) 1969–1971.
      b. B. Schmidt, D. Geißler, Eur. J. Org. Chem. (2011) 4814–4822.
      c. B. Schmidt, D. Geißler, Eur. J. Org. Chem. (2011) 7140–7147.
      d. T.J. Donohoe, L.P. Fishlock, A.R. Lacy, P.A. Procopiou, Org. Lett. 9 (2007) 953–956.
      e. Y.K. Yang, J. Choi, J. Tae, J. Org. Chem. 70 (2005) 6995–6998.

CHAPTER *6*

Summary and Outlook

The achievements of furan synthesis based on transition metal catalyst have been summarized and discussed. The whole manuscript is divined by the type of the reactions. Hopefully, it will be useful for our community.

As expected, there must have been mistakes from place to place. I'm here asking for your forgiveness for this and also for the possible missed literatures.

Transition Metal Catalyzed Furan Synthesis. DOI: http://dx.doi.org/10.1016/B978-0-12-804034-8.00006-4

Printed in the United States
By Bookmasters